❖

THE CHANGING BRAIN

ALZHEIMER'S DISEASE AND ADVANCES IN NEUROSCIENCE

❖

IRA B. BLACK

OXFORD
UNIVERSITY PRESS
2002

OXFORD
UNIVERSITY PRESS

Oxford New York
Auckland Bangkok Buenos Aires
Cape Town Chennai Dar es Salaam Delhi Hong Kong Istanbul
Karachi Kolkata Kuala Lumpur Madrid Melbourne Mexico City Mumbai
Nairobi São Paulo Shanghai Taipei Tokyo Toronto

Library of Congress Cataloging-in-Publication Data

Black, Ira B.
The changing brain: Alzheimer's disease and advances in neuroscience / by Ira B. Black
p. cm.
Originally published: New York McGraw-Hill, 2001.
Includes bibliographical references and index.
ISBN: 0-19-515697-8 (pbk.)
1. Brain—Popular works. 2. Neurosciences—Popular works.
3. Alzheimer's disease—Popular works. I. Title.

QP376.B6235 2002
612.8'2—dc21 2001058378

1 3 5 7 9 10 8 6 4 2
Printed in the United States of America

CONTENTS

Contents

PREFACE:
THE CONSTANCY OF CHANGE

Few of us are indifferent to changes. We greet a new house, a new spouse, a new child, or a new job with excitement, anxiety, dread, or euphoria, but rarely with indifference. Indeed, our species celebrates change with enchanting, elaborate ceremonies. Birth, growth, adolescence, marriage, and death are guideposts that shape and direct our lives. Our passion for change is directed to the outer world as well. We create imaginative holidays to mark the new year, the rebirth of spring, the summer's solstice and growth, the autumnal harvest, and the frigid winter of laying away. But few of us pay attention to the continuous internal changes that are the substance of our biological heritage, that unite us with all life, that endow us with our humanity.

Biological change is part of the larger process of change in the universe. The world around us is not static, not unchanging, but evolving from the shattering instant of the big bang to a much-debated future. From the first three minutes to the last, the universe is in constant flux, cooling profoundly, moving through phase transitions, creating atoms from fundamental particles, forming heavier elements from primordial hydrogen, building molecules from atoms, condensing matter that accretes galaxies, clusters of galaxies, and stars everywhere. With the improbable appearance of life, or perhaps with the survival of one life form after abortive appearances, the world witnessed novel changes.

Changes of life occur at multiple levels in time and space. Grand changes in evolutionary time transpire at a glacial pace, nearly invisible to our brief, four-score glance. Over billions of years of evolutionary time, the biosphere, clades, species, individuals, traits, cells, and genes adapt to a changing environment. Over thousands of generations, environmental conditions select among competing individuals. They favor the reproduction of those best adapted and enable the chosen to pass on traits suited to the environment of the time. Relentless choosing, reproduction, and rechoosing yield an exquisite fit of organism and environment. One primary catalyst for evolution is development.

The changes of development adapt the organism to the environmental demands, internal and external, of each stage of life. From conception to birth to reproduction, selective pressures favor the best suited individuals at each stage. Thus the processes of development become the nuts and bolts, the bricks and mortar, on which selection works. Biological systems effectively pick and choose among developmental mechanisms, combining new processes to adapt and change throughout life. The alphabet of mechanisms is introduced during ontogeny, and biology constructs new words, sentences, and paragraphs to grow with experience and to pass on to successive generations. In this way, our evolutionary history, and the environment in which our forbears evolved, continue to shape the very mechanisms of our changing lives. Evolution and natural selection are not disembodied abstractions. They hold the keys to our nature.

❖

Who are we? Where did we come from? Are we mind or brain? How does the organ of instant thought and emotion also store experiences for decades? Recent discoveries are providing answers. Neurons send electrical impulses in thousandths of a second, hold electrical conversations for seconds to minutes, but grow over days and years. The influence of moment-to-moment electrical signaling on long-term growth is central. Experience creates fleeting electrical impulses; the impulses elicit growth; the growth stores information.

Some of the roles of electrical impulses have been known for decades. When we see a sunset, hear a song, or smell a rose, neurons fire impulses and record the experience. Electrical signals jump the gap to the next neuron in line, relaying the information. The junction between neurons, the synapse, is critical in regulating information flow. If the synapse is strong, more information is relayed; if it's weak, less is communicated. The more efficient the synapses in a system, the better the information is transferred.

Our view of synapses is changing radically. We now know that synapses are not unchanging, hardwired, switchboard connections, as previously assumed. They continually strengthen, weaken, grow, and shrink throughout the brain's life. Experience itself regulates the growth and strength of synapses.

Growth is not restricted to synapses, however. The whole neuron is ever-changing, growing, and remodeling, autonomously and in response to the environment. Growth is central to normal brain function. Growth changes the structure and function of neurons, the synaptic connections among neurons, and the architecture of networks of neurons. In short, it is key to the understanding of brain and mind. How does growth occur? Can we identify growth mechanisms?

Contrary to traditional teaching, growth is not simply governed by predetermined, immutable internal programs. Growth is also experience-driven. Brain function represents a balance between innate and experience-driven growth. Brain growth results from cooperation between nature and nurture. In a broader sense, growth is productive, creating new biological contexts within which genes and cells operate and develop. Growth creates its own new environments, its own new experiences. Consequently, there are no easy divides between nature and nurture, genes and environment, biology and everything else.

Experience accesses the growing neurons through special signal molecules, which can be divided into three families. Neurotransmitters relay moment-to-moment electrical impulses between neurons. Growth factors cause neuronal cell division. Trophic factors support neuronal survival. Yet all three classes of molecules—neurotransmitters, growth and trophic factors—also cause neuronal growth. This realization prompts us to focus on molecules and neurons as units of information processing.

The submicroscopic world of molecules possesses reasoning power, or computational power, that is thousands of times greater than it is at the neuronal level. For example, a single enzyme molecule performs 1 million chemical reactions per second. In contrast, whole neurons convey impulses in thousandths of a second. The enzyme wins. It conducts its business a thousand times faster than the neuron. Molecules are also convenient—they don't take up much space, being only one-hundredth as long as a garden variety synapse. Converting this one-dimensional difference of 100 to three-dimensional space ($100 \times 100 \times 100$) yields a difference of 1 million; hence molecules pack a millionfold more computational power into space.

At every level of organization, from molecule to neuron to neural system to networks of systems, the process of growth distinguishes the brain from other reasoning gadgets. Abacuses, calculators, and computers don't grow. Yet what is the substance of growth? What actually happens? We take a broad view of the scope of growth: extant molecules are modified, new molecules form, molecules aggregate to create new cell structures, neurons make new connections. These are only a few of the processes the brain uses to incorporate change. Destructive, or regressive, growth events are equally important. Removal of molecules, remodeling of cell structures, weakening of neuronal connections, and even cell death are also used by the brain to encode information. Growth, then, is a strategy unique to life, enabling us to learn, remember, think, feel, and be human.

This capacity for change is built into the brain at multiple levels, and is also termed "plasticity." Plasticity is innate; it obeys the rules of biology. But individual plastic processes can be activated by the environment. In a paradoxical way, plasticity, which confers flexibility, follows the strict rules of biology; it adheres to inflexible rules to achieve brain flexibility.

Plastic growth mechanisms draw on transmitter, growth, and trophic factor molecules. Neurons use transmitter and trophic factor communication to accomplish a remarkable feat: the conversion of momentary environmental events into neuronal changes lasting months to years. This is how it works: Brief experiences touch off neural impulses that release transmitter signals. The transmitter jumps the synaptic gap to electrically stimulate the next neuron in line. The transmitter activates genes in the next neuron that make trophic factor, and this factor strengthens the synaptic connection. Synaptic strengthening can then last for weeks, at the least. Thus plasticity allows brief experiences to strengthen a neural pathway over the long term.

Growth mechanisms allow the brain to capture time. Brief experiences change the structure of synapses, and the changes persist over time, creating memories. Memory mechanisms are built into the rules governing brain architecture. Changes in the environment translate into changes in the brain. Plasticity allows us to embody time. We are not doomed to live in an eternal present; we are beings with a history. And awareness of history, made possible by memory, permits learning.

When learning and memory fail, we confront baffling and frightening diseases. Alzheimer's disease, for example, robs the brain of miraculous capacities and robs the patient of his or her humanity. With our new understanding of brain function and growth mechanisms, though, we can frame questions in an educated way. Brain neurons that form memories die in Alzheimer's patients. Is Alzheimer's due to deficits in life-sustaining trophic factors? Is the dementia due to abnormalities in transmitter signaling? Can the new factors be used as treatments to prevent the death of brain neurons? Our knowledge of growth mechanisms and molecules helps in the search for answers.

❖

How can we examine genes, brain, and mind in a single brief book? We experiment. This book itself is an experiment. To approach the most complex system in the biological universe we tell two parallel but related stories. We live with Enoch Wallace, the successful investment banker who progresses from imperceptible distraction to momentary mental lapses to transient memory deficits to disorientation, dishevelment, emotional lability, incontinence, and institutionalization. We experience his uncertainty, confusion, anxiety, and dread as Alzheimer's disease destroys his brain, mind, and humanity.

Enoch's tragedy is viewed in the context of a second story, the revolutionary discoveries of modern neuroscience. While multiple motivations prompted simultaneous presentation of the twin tales, two objectives were paramount. Nature's tragic experiment of disease helps breathe humanity into the science of molecules, neurons, and brains that otherwise may remain remote. Hopefully, the details of Enoch's symptoms and suffering endow the abstractions of brain science with human meaning.

Our second objective in telling the clinical and scientific tales in tandem is unabashedly pedagogical. Complex, sometimes obscure neuroscientific concepts often become graphic and accessible when viewed through the prism of dysfunction and disease. The opaque logic of laboratory experiments can gain transparency when we understand their relevance to our own brain function. The disintegration of Enoch's brain illustrates underlying principles of function that can elude conventional neuroscientific description.

This inspired plan is not without risk. Two stories can be more than twice as confounding as one. Threads and themes have a way of self-organizing into Gordian knots. Injudicious admixture can create a chimera that confuses layman, scientist, and clinician in equal measure. To help avoid this unhappy circumstance, the two accounts are presented separately in each chapter. I begin by describing Enoch and the progress of his disease. Enoch's episodes are followed by broadly relevant neuroscientific discussions, though there is not a strict one-to-one correspondence be-tween Enoch's experience and the brain science in each chapter. Rather, the two stories progress in parallel through the book with hoped-for emerging coherence. Enoch's episodes are followed by an additional brief clinical vignette that is directly relevant to the neuroscientific dis-cussion in each chapter to help us move from bedside to laboratory.

The strategy is designed to indicate how laboratory and bedside are different arenas of an integrated campaign. Meticulous observation of pa-tients with brain disorders has led to deep insights into mind function, leading in turn to novel laboratory experiments resulting in even deeper understanding. Conversely, basic science has defined disease mechanisms and processes. Neuroscientific discoveries are now leading to treatments inconceivable even ten years ago. And the integrated enterprise is pro-viding a new vision of brain, mind, and human nature.

Ira Black
Robert Wood Johnson Medical School
New Jersey

❖

ACKNOWLEDGMENTS

Mike Gazzaniga has long encouraged students and colleagues to forsake the security of their cloistered areas of expertise and synthesize across foreign areas, though such risky ventures are often greeted by discomfort, twitters, snickers, and opprobrium. These novel forays, however, can lead to discovery and insight. Mike brings the same sense of supportive, goading encouragement to the alien adventure of writing for a non-technical audience. For many scientists, this one included, a journey beyond the secure confines of impenetrable facts, figures, citations, and jargon can elicit deep anxiety. I thank Mike as much for his optimistic sense of adventure as for his critiques of manuscript drafts.

I also thank Mike for introducing me to Howard Boyer. Friend, colleague, and guide, Howard commented on chapter after chapter, insisting on clarity in a medium approaching English, protesting as details led to terminal boredom, and exhibiting a feel for the written organism that was edifying.

Michael Lewis at Robert Wood Johnson Medical School helped minimize inconsistencies, redundancies, and ambiguities in early drafts, though no effort could be entirely rehabilitative. Seamless discussions during many walks in the parks of Princeton helped crystallize inchoate ideas and concepts. Mike also modeled an emancipating style that contributed to progress immeasurably.

Reed Black reviewed many chapters with a wonderful eye for my ability to be simultaneously trivial and incomprehensible. He also identified the pockets of interest and significance that required amplification.

Janet Davis was ceaselessly encouraging, evaluating through the clarifying prisms of organizational psychology, social systems, and the process of publishing itself.

Acknowledgments

Piergiorgio Strata generously located several critical publications and provided alternative translations from the Italian. His conscientiousness is gratefully acknowledged.

Friends and colleagues, principally including Dick Goodwin, Steve Pinker, Daphné Ireland, John Tooby, and Leda Cosmides, unwittingly helped with seminal discussions at various stages.

Betty Wheeler offered expert organizational and technical guidance, allowing the work to actually achieve a physical existence.

Amy Murphy orchestrated and edited with a delicate, sagacious hand, happily integrating enthusiasm and discipline.

Colleagues at McGraw-Hill, including Scott Amerman, Karen Auerbach, Ede Dreikurs, and Lynda Luppino, forged an ever-helpful, effective model team.

Our laboratory work has received the generous support of a number of agencies and foundations, that indirectly fostered this book. The National Institutes of Health made possible an extraordinary number of the described advances. The success of the NIH, and its underlying mission, represent the best that we are. In particular, I thank Gilman Grave, National Institute of Child Health and Development, for ongoing encouragement and support. The Christopher Reeve Paralysis Foundation to cure paralysis has been inspirational. I also thank the McKnight Foundation, The National Foundation-March of Dimes, The Unrestricted Award Program of Bristol Myers-Squibb, and Juvenile Diabetes Foundation for support in the past.

THE CHANGING BRAIN

❖

*Enoch Wallace, investment banker, experiences his first,
frightening momentary lapses of memory. Nerve
connections deep in his brain are beginning to
malfunction in early stage Alzheimer's disease.
Communicative connections among brain neurons
generate mind, emotions, and our humanity. Special
signal molecules from targets attract nerve processes that
form the all-critical connections constituting brain and
mind architecture. The discoveries leading to these
insights take us from Italy to England and Switzerland.*

❖

1

THE BRAIN CONNECTION

❖

ENOCH WALLACE, a distinguished sixty-two-year-old investment banker, awoke at 6:00 A.M, looking forward to getting to his office on Wall Street and closing the deal on a complex corporate acquisition. He quietly sat up in bed, as he had for twenty years in his Park Avenue apartment, to avoid disturbing his wife. He stared across the room, taking in the familiar Winslow Homer seascape hanging on the opposite wall. As he rose to begin the morning ritual of preparation, his familiar world changed utterly. He could not remember how to get to the bathroom. He couldn't navigate a route that he had negotiated thousands of times. After the disbelief and then the denial that anything was amiss, he panicked. Then, suddenly, the clouds parted. He walked to the bathroom as if nothing had happened. He washed his face with the "Scent of Irish Mist" soap, brushed his teeth from the half-filled Crest tube, and showered uneventfully. The driver picked him up at 8:00 A.M., and he was in the conference room by 9:00. The day of negotiations, phone calls, electronic mail, mad huddles from room to room plotting bargaining ploys, and threats and cajoling, proceeded as planned. Enoch reveled in the ordered freneticism, the disciplined disputations of another merger in progress. He left "the street" with the satisfaction that a hyperkinetic day, devoid of distracting reflection, can bring.

He was in an unaccountably celebratory mood, and looked forward to a smooth, dry martini after stopping at the florist for Sally's favorite yellow, spiked mums. The driver dropped him off in front of B&D Florists, and disappeared into the hail of rush hour traffic. He walked in, said hello to Joe, and headed for the stem flower collection. On leaving the store, before any conscious awareness, he was gripped by the forgotten, frightening confusion of the morning. Though only two blocks from home, on a route he had traveled for ten years, he didn't know which way to turn. He recognized the stationery store, the small restaurant across the street, the gourmet butcher, but he was lost. His heart was racing again, he was short of breath, his mouth was dry; he turned back to B&D, and joked with Joe about the shortest way back to his apartment. Joe laughed, said that "you must'a had a busy day closing deals," and guided Enoch to the shortcut behind the store. Enoch finally stood in front of his apartment door with a bewildered sense of confusion and relief. What exactly had happened? Was today's rush of negotiations particularly arduous and straining? Had he been working much harder than he thought? Should he take that vacation that Charlie had been recommending? He had never been absentminded. He brushed it all aside, entered the apartment, experienced mild relief on seeing the familiar surroundings of the Queen Bokhara carpet, the octagonal dining room table, and the hutch, and began cutting the flowers for the Finnish cut glass vase. The smooth martini was on his mind.

Enoch Wallace was experiencing the loss of spatial memory that is an early hallmark of Alzheimer's disease. Neurons deep in his brain were not functioning normally.

❖

Laurie Grundfest, forty-seven years of age, took the same bus home from teaching the third grade for fifteen years. On a sunny fall day, she got off at her stop to walk the five blocks home. On this day, for the first time ever, she could not remember the way home. Though the stores, trees, and traffic lights were familiar, she did not know which way to turn. She approached a policeman, told him her address, and he escorted her home. Her early stage Alzheimer's disease was attributable to the degeneration of a group of nerves and their connections in her brain. Within seven years she couldn't remember the names of her children.

How does faulty neuronal communication resulting from compromised connections interfere with memory? While partial answers will emerge in our discussions, examination of an entirely different clinical case may help relate connections, communication, and brain organization.

❖

Charlene Washington is a sixty-two-year-old right-handed grandmother who works as a Human Resources supervisor in a Pennsylvania pharmaceutical company. She has a twenty-year history of hypertension, which she has treated irregularly. Charlene has suffered several strokes. The most serious occurred two years ago and interferes with vision on her left side. However, she has learned to compensate, and her warm, sympathetic personality and people skills allow her to perform her supervisory roles superbly. Her visual impairment certainly has not compromised her close relationships with her daughter, Becky, or with her two young grandchildren, Sam and Kenny.

On this Tuesday morning in June, she sat at her desk discussing a personnel problem with Gary, the labor relations representative. As Charlene took notes, she was startled when she suddenly dropped her pen. She reached to retrieve the ball-point, but her fingers had become awkward and weak and she could not manipulate them. Her right arm felt heavy, and she experienced difficulty reaching across her desk. She attempted to relate her confused surprise to Gary, but her speech was strangely stuttering, hesitant, and garbled. Alarmed, Gary called security, and Charlene was rushed to the local hospital in an ambulance.

In the emergency room, and then in the intensive care unit, she was placed on intravenous medication. Charlene was experiencing an evolving stroke. A blood clot was clogging a critical brain artery. Although she maintained alertness throughout, her neurological deficits followed an up-and-down course over the next six hours. Her speech improved, but then deteriorated again several times. The weakness in her right hand and arm waxed and waned, varying from near-total paralysis to a mild sense of heaviness. On several occasions Charlene had difficulty moving her right leg, but this never persisted and finally relented altogether. At no time did she experience pain. Ten hours after initially dropping the pen her condition stabilized miraculously. Her speech returned, she had no weakness in her hand or arm, and her leg was fully mobile. However, meticulous testing by a neuropsychologist revealed a bizarre language disorder.

When Charlene was shown a pen, a comb, or a book visually, or anything else for that matter, she could not name the object. Yet, if she was allowed to touch and feel the object, she named it rapidly and normally! Even if hidden from her view behind a screen, she could name a penny, a dime, or a quarter accurately as long as she was able to explore the coin with either hand. However, when merely shown the coin, and not allowed to touch it, she was thoroughly incapable of identifying it. Charlene was suffering from a brain "disconnection syndrome." A brief explanation will illustrate the centrality of connections and communication in brain function, and will help as we strive to understand the Alzheimer deficits in Enoch.

Naming a visually perceived object requires the visual areas of the brain, which receive images, to send impulses to the speech area. The speech area can then perform the naming. But in Charlene's brain, the visual areas are disconnected from the speech area. There are two visual areas in the brain, one on the left and one on the right; the left visual area receives images from the right side of the field of vision, and the right visual area from the left side. Charlene's right brain visual area was damaged in a previous stroke, causing a left, opposite-sided visual deficit. The new stroke severed the connections from the remaining left visual area to the speech center. Consequently, visual information could not reach the speech area by any route, and Charlene was prevented from using speech to name visual material. However, the touch centers were left intact and remained connected to the language center. Information flowed normally from the touch centers to the speech area, allowing her to name objects that she could touch and feel.

The separation of the brain into discrete centers, each performing specialized functions, and their need to connect and communicate, is a basic principle of brain function. Depending on which centers are disconnected, disorders may occur in language, motor function or cognition, including memory.

❖

After centuries of often bloody debate about body and soul and the nature of man, we now understand that the brain's architecture gives rise to mind, feeling, and personality. Our brains endow us with our humanity and determine who we are as individuals. The tragedies of stroke, Alzheimer's disease, and head injury, accompanied as they are by dementia and personality fragmentation, bear undeniable witness to brain as the basis of mind. But if brain structure defines mind, what exactly changes when we learn and remember? Our challenge is to discover how experience changes the brain and hence the mind. As we understand more about brain changes, we hope to one day prevent senility, depression, anger, and hate. We already know enough to improve learning and memory in everyday life, and to do a far better job of realizing our potential; conversely, our new insights are prompting novel approaches to dreaded brain diseases.

In a system of one trillion neurons and thousands of circuits, which structures generate mentality, emotion, and individuality? How do changes in these structures lead to a memory, the acquisition of a new skill, or a change in personality? Signaling connections between neurons are, we know, central to thinking and feeling; these connections transmit information from one neuron to another, converting a single dumb neuron into an intelligent, conscious community of neurons. Somehow changes in these all-important connections affect learning, memory, and personality.

Brain Signals

To figure out how neuronal connections change, we have to know something about mechanisms of action. A neuron communicates by sending a chemical signal across a synapse (Figure 1-1). It releases the chemical in response to an electrical impulse. The signal then moves across the connection and stimulates the receiving neuron. The receiving neuron converts the chemical signal back into an electrical impulse before beginning the process again and communicating with the next neuron in line. In this way, neurons hold electrochemical conversations over thousandths of a second, enough time to evoke an instantaneous thought, a fleeting emotion, a muscle movement, or a smell, depending on which neural system is active.

Moment-to-moment signaling does not ordinarily change the structure of the connection; it simply transmits the signal passively, like a telephone, without itself changing. For years scientists had imagined that the connections were complex but stable signaling devices that faithfully relayed information in a slavish fashion. They regarded the whole brain as a frightfully complex but fixed switchboard.

Two seemingly unrelated discoveries have revolutionized our views. They hint that the brain is no less dynamic and no less changing

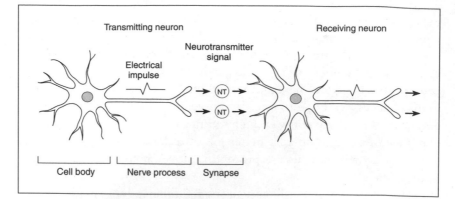

Figure 1-1   Communication at the synapse. In response to an electrical impulse, the transmitting neuron releases neurotransmitter (NT) signals that stimulate electrical impulses in the receiving neuron. In turn, the receiving neuron activates the next neuron in line.

than our very thoughts and emotions. After years of experiments, researchers have recently discovered that neuronal connections do indeed change with cellular experience. Scientists examining a part of the brain critical for memory were surprised to find that rapid electrical stimulation heightened the neuronal connections' efficiency.[1] A burst of stimulation lasting only a few seconds strengthened the connections for hours. The persistence of change long after the stimulation ceased, looked suspiciously like memory. To add to the intrigue, a drug that hindered strengthening prevented rats from remembering how to navigate in a maze for a food reward.[2] Truly significant relations among brain electrical activity, connection efficiency, and memory were coming to the fore.

Growth Signals

The other revolutionary discovery occurred in an area that seemed remote from the brain. Rita Levi-Montalcini, working secretly in her bedroom fifty years ago in fascist Italy, and later as a visiting scientist at Washington University in St. Louis, discovered a hormone that induced certain nerves to grow. In tissue cultures of nerves, and in a live animal, the hormone prompted nerves to grow in size and number, and to send out long fibers. The hormone was, not surprisingly, named nerve growth factor, or NGF for short.[3] The responsive nerves that Levi-Montalcini studied were located outside of the brain, but her discovery inspired scientists throughout the world to search for similar effects on brain neurons. NGF was administered to animals and added to tissue cultures in an endless variety of experiments, but the brain neurons did not respond. Obviously they couldn't examine a trillion neurons, so scientists focused on brain neurons in the same chemical family as the ones that Levi-Montalcini stimulated with NGF. The failure to stimulate brain nerves with NGF reinforced the dogma that brain neurons simply don't grow after development. Efforts to study growth were largely abandoned as hopeless.

Within the past decade, however, a new generation of biologists has come of age. Equipped with new, more sensitive methods to grow brain neurons in pure culture, this new breed has not felt bound to honor old dogmas. Also during this time, neurologic diseases that had been hopeless mysteries through the centuries began to yield to the reinvigorated neuroscience. A major group of neurons that degenerate in Alzheimer's disease was discovered deep in the brain. The new scientists were able to grow these neurons in the well-controlled environment of the dish and thus define what was needed for the neuron to thrive—and what might cause their death. Employing a judicious mixture of wishful thinking, intuition, insight, and desperation, several laboratories in Europe and the United States simultaneously came up with the idea of adding Levi-Montalcini's NGF to the cultured "Alzheimer neurons." The results were astonishing. NGF dramatically prolonged the neurons' survival. Disbelieving experimenters repeated the tests in endless variation with the same result: NGF extended the life of neurons that normally die in Alzheimer's patients.[4]

The flood gates opened. Within months, new growth factors were discovered and cloned. Old growth factors, which reside elsewhere in the body, were also detected in the brain. Entire families of growth factors were discovered, and, almost monthly, yet another population of brain neurons was seen to depend on a certain factor for growth and survival. Almost overnight the brain was transformed from a static, immutable, hardwired switchboard into a vast reservoir of growth factors that directed growth, survival, and previously unimaginable flexibility.[5]

Surprising relationships among growth factors, connectivity, neural communication, and brain function are now emerging. Growth factors are critical brain signals that strengthen the neural connections lying at the heart of brain function.[6] Brain growth factors convert experience into strengthened neuronal connections that mediate learning, memory, and emotion. That immodest statement requires explanation.

The implications are profound: the brain is a growing, ever-changing, flexible structure that generates our plastic minds. How, exactly, does experience change the brain and result in learning and memory? To answer this, we have to first identify important environmental events and how we experience them. Then we have to describe how we experience the world, how we learn from the experience, and how we optimize learning. Once we have developed this perspective, we can delve into underlying brain mechanisms. Let's consider first some salient aspects of experience and how it generates memories.

Experience Shapes Learning

Internal change begins when a new external experience activates brain pathways. The external world of unanticipated experiences seems beyond our control, but scientific insights may enable us to design our experiences to realize our potential. What, indeed, does that mean? A high school student memorizes the achievements of manned space flight for her world history test. What is the best way for her to study in order to maximize the neural activity that strengthens her neuronal connections? We don't know, alas, but significant hints have accumulated over the past century. The famous psychologist Hermann Ebbinghaus noted that learning is most effective when the study time is spread over days rather than crammed into a single session the night before the test. Translating his turn of the twentieth century observation into twenty-first century neuroscience, we might guess that study spread over time is most effective at strengthening neuronal connections.

How else might we shape experience to optimize mind changes? Fascinating experiments provide additional clues. Students learned best when their studying was followed by a night's sleep. Sleep itself was the critical factor, leading to further examination of the potential connections among experience, learning, and sleep. Experimental studies of rats provided important insights. Rats were taught to navigate through a complicated maze while the electrical activity of their brain neurons was monitored. The particular neurons examined were of special interest, since they were part of the brain known as the hippocampus. This brain region plays a crucial role in memory, and is especially important for spatial memory, such as maze learning. As the rats learned to navigate the maze successfully, the hippocampal neurons fired with a characteristic, unique pattern. This was remarkable in itself: a special pattern of nerve activity was associated with learning a particular task. But the truly astonishing part of the experiment was yet to come: during sleep, the hippocampus reproduced the learning firing pattern.[7] Was the hippocampus practicing the learning pattern; did pattern rehearsal strengthen neural connections, consolidating learning and memory? These questions are now under intense examination, but we can say at this point that the foregoing results counsel us to think, and then sleep on it.

One final example illustrates how experience and psychological state markedly affect neural connections and learning. Our liberated educational system tends to view stress as deleterious and favors a more laid-back approach to learning. The neuroscience laboratory now suggests that life is more complicated, and that a little stress may do wonders for neuronal connections and learning. Once again, maze learning in rats provided surprising insights. One group of rats was stressed with a mild tail electrical shock, and the control group received identical treatment without the tail shock. The group that received the shock stressor learned faster, and retained the knowledge longer. Moreover, in this case, the neuronal connections were studied. The stressed rats strengthened neuronal connections more efficiently.[8] So, at least some degree of stress in the environment appears to aid some forms of learning. We could speculate that stress somehow enhances alertness and attention, brain states that foster connectional strengthening and learning.

These few examples illustrate that we can begin to optimize strategies for learning, and, by implication, for teaching, rooted in neuroscientific principles. We may prescribe a moderate dose of pretest stressful jitters to motivate the student, the worker facing a new task, or the manager designing a new project; we may recommend spreading out the thinking process over a number of days, and getting several nights of restful, rehearsal sleep before attacking the challenge at hand.

More generally, we are not simply passive objects, at the mercy of life's challenges. We can structure the environment, the style of our experiences, and our participation as stressed or somnolent beings, to fully utilize our capacities. In sum, before tweaking esoteric, recently discovered brain mechanisms to expand mind, we can style the environment and better prepare ourselves to enrich experience.

In addition to providing new guidelines for optimizing learning and memory, the brain rules uncovered by neuroscience provide entirely different ways of thinking about the mind. We often think of a friend's memory abilities as a hard-wired, fixed part of his brain. "Samantha has a good memory, but Bill's memory is terrible." Our newfound insights suggest that mind and brain are not that simple, and that memory abilities are subject to change. Perhaps Bill has not "learned" how to use a little stress to sharpen his "memory." Perhaps Bill's insomnia interferes with normal hippocampal rehearsal, while Samantha gets the full night's sleep that consolidates memories. Maybe Bill's memory problems would vanish with increased exercise, greater fatigue at night after studying, and improved sleep habits.

The Growth of Neuronal Connections

Our central problem is to understand how environmental challenges, whether navigating mazes, apartments, or the few blocks home, are solved by the brain in the form of learning and memory. The brain contains one trillion neurons and one million billion (10^{15}) neuronal connections. We must look at the synapses, the neuronal connections that transmit information, for clues to the puzzle of learning and memory. We have seen that the discovery of brain growth factors has pointed to a potential link between experience and the alteration of synapse efficiency. To approach this emerging relationship, it is helpful to describe the history of NGF (nerve growth factor).

Levi-Montalcini's experiments of fifty years ago revealed that NGF had profound growth effects on neurons lying outside the brain.[3] Addition of NGF to culture dishes containing groups of neurons resulted in striking changes. Peering through the microscope, she observed marked increases in the sizes of the neuronal groups. Both the number of neurons and the size of each individual neuron increased dramatically. But NGF also evoked an even more startling change. The hormone caused the neurons to send out long fibers. In a few days, the fibers virtually covered the culture dish. These are the fibers that neurons use to contact each other. These fibers develop specialized contacts to form synapses.

When Levi-Montalcini treated live rats and mice with NGF, she saw similar effects, indicating that she was not simply observing a peculiarity of cultured neurons. Again, NGF treatment increased the size and number of neurons lying outside of the brain. It also caused a stunning increase in the number of nerve fibers in target organs of the neurons, such as the heart and iris, which are controlled by the nerves. NGF appeared to be increasing the connection of nerves and their targets in living animals.

Since administered NGF had such dramatic actions in the animals, it was natural to ask whether NGF was normally present in the body, and whether it normally regulated nerve function. However, there were no reliable methods to measure NGF directly. Levi-Montalcini used an indirect approach. She administered an NGF antiserum, which could neutralize any NGF present in the animals, and prevent actions on neurons. The results were again dramatic. The NGF-neutralizing antiserum caused the death of many neurons and essentially destroyed all the fiber connections of the neurons with their targets. A number of tentative conclusions were drawn.

The antiserum experiments suggested indirectly that NGF was, in fact, normally present in the body, somehow affecting neurons. The antiserum and the tissue culture experiments further suggested that NGF was necessary for the survival of certain neurons. Finally, and most importantly, NGF appeared to cause nerve fiber growth, and the connection of nerves with their targets. As so often happens in science, these answers raised even more baffling questions. How could NGF cause the growth of nerve fibers, and mysteriously guide the fibers to far distant targets to make vital connections?

The Cambridge Connection

This was the state of knowledge, and ignorance, as the turbulent 1960s drew to a close. The scene shifts to Cambridge, England, a storied world center of science, scholarship, and discovery. In 1970, the University of Cambridge was still basking in the afterglow of the Watson-Crick discovery of the structure and significance of DNA, the basis of life. Many of the architects of molecular biology had already departed for an even newer frontier, the new science of the brain, neuroscience. Francis Crick remained at Cambridge, contemplating consciousness and perception. Sidney Brenner, also at Cambridge, launched a campaign to define the nervous system of a simple animal, a worm, in its entirety, down to the finest detail, in an attempt to pierce to the essence of brain function. Seymour Benzer, another of the molecular biology pioneers, chose the fruit fly, long a model for biologists, to elucidate the genetics of neural function in his laboratory across the Atlantic, at Caltech.

One of the jewels of University of Cambridge was Trinity College, the home of Sir Isaac Newton. A brilliant thirty-five-year-old neuroscientist, Leslie Iversen, had recently established his laboratory at Trinity, and students were beginning the traditional international scientific pilgrimage to train and collaborate with him. The laboratory consisted of two postdoctoral students, two graduate students, two technicians, and Iversen. When a graduate student himself, Iversen had discovered how a class of neural signals, neurotransmitters, conduct some of their operations, and had written the authoritative book on one of them, noradrenaline.[9] He had recently returned from a year working in the laboratory of Julius Axelrod, a future Nobel Laureate, at the National Institutes of Health in Bethesda, Maryland.

Though modest in size by international standards, Iversen's laboratory was already international in flavor, with an American and an Australian among the British. Ian Hendry was a doctoral student who had obtained his medical degree in Canberra, Australia; his busy schedule included religious attendance at pubs, conscientious sampling of British beer, rugby, and experiments on NGF. Hendry sported an apparently casual air, which masked some deep thinking about NGF and the nervous system. He had a way of jokingly speculating about a scientific point in passing and then proving it with hard-edged experiments.

As part of his doctoral project, Hendry was devising a new method to reliably measure NGF in the body. One of his objectives was to find out where the growth factor was localized, and thereby begin to understand how it works. Levi-Montalcini's antiserum experiments were very much on his mind. He was wrestling with apparently unrelated facts: nerves extend long fibers toward their targets; the fibers enter the target tissues and form synaptic connections with target cells; NGF somehow dramatically increases growth of fibers; NGF markedly elevates the number of nerve fibers in targets; NGF is necessary for the survival of the family of neurons in question. The problem was how to relate neuronal survival, fiber outgrowth, target contact, the formation of connections, and NGF itself.

Somewhere between pub hops, the rugby field, and the laboratory, Hendry came up with a wild hypothesis to put the pieces of the puzzle together. He speculated that nerve targets produce NGF, which the nerves require to survive, and which attracts nerve fiber growth. In one imaginative fell swoop this hypothesis would account for many of the puzzling relations between nerves and their targets. Over the next years and decades, this hypothesis and its progeny would answer questions that had not yet been asked.

A successful hypothesis tells a story of a sort, one that takes multiple odd characters, strange events, unfamiliar settings, unlikely circumstances, and weaves them into a compelling narrative. The (scientific) audience is left breathless with the sense that it must be true, since it makes order of the chaos of characters, events, settings, and circumstances that are otherwise unintelligible. The particularly elegant hypotheses accomplish all this with uncommon economy. The so-called neurotrophic hypothesis is of this sort. Hendry speculated that targets produce the NGF that is required for neuronal survival. During development, before any nerves have sent fibers to targets, each individual target releases NGF like a radio transmitter, sending forth waves in all directions. In response, distant, spherical nerve cell bodies extend fibers that grow upstream to regions of ever higher concentrations of NGF, until the target is reached. Once the nerve fiber has reached the target, and made the all-important connection, it is assured of obtaining NGF and surviving. This simple story accounts for many peculiarities of the nervous system.

In this scenario, nerves grow toward targets initially because they are attracted by the life-giving NGF. Those nerves whose fibers succeed in reaching the target live by virtue of contacting a source of NGF. Nerves that do not find the target die of NGF deprivation. In fact, this is precisely what happens during development: the majority of neurons die, since they lose in the competition for NGF. Also, this elegant mechanism ensures that precise, point-to-point circuits are accurately formed in the nervous system that owes its function to selective communication along specific pathways. The hypothesis, consequently, explains mechanisms underlying the formation and maintenance of connections in the nervous system. If these processes were also detectable in the brain, then we would understand a great deal about the connections generating learning and memory.

The last "if" is a big one. Like any good hypothesis, this one raised yet other unanswered questions. Hendry, Iversen, Levi-Montalcini, and others were working only with peripheral neurons that lie outside of the brain. For the moment, no one was thinking about application of the NGF target hypothesis to brain connections. Yet the brain has hundreds to thousands of different specific pathways. Surely, a single signal such as NGF could not account for the specificity of hook up of all these neurons and targets simultaneously. Buried deep within this hypothesis was the probability that many other growth factors were lurking in the nervous system. This did not trouble scientists at the time. They had enough work to do to prove or falsify this hypothesis.

A flurry of experiments designed to test the NGF target hypothesis rapidly followed.[10] Hendry and Iversen initially took a direct approach. They reasoned that if peripheral neurons depend on target NGF for survival, separation of neurons from their targets should lead to death. Hendry surgically cut the fibers that connect the nerve cell bodies with distant targets. He used young rats, since Levi-Montalcini's experiments with anti-NGF antiserum had shown that their neurons were particularly dependent on NGF. As predicted, neurons that were surgically separated from targets shrunk and died. So, targets were critical for the survival of neurons.

To determine whether NGF and not some other factor was the important target survival agent, Hendry added an experimental wrinkle. In one group of animals, he cut the fiber connections with the targets, but treated the rats with NGF at the same time. NGF could reach the neurons through the circulation, bypassing the need to derive the factor from the target. NGF treatment dramatically prevented the death of neurons separated from their targets due to fiber connection cutting.

These simple experiments led to several important tentative conclusions. The results indirectly suggested that NGF was produced by neuron targets. The results also substantiated Levi-Montalcini's contention that NGF is necessary for the survival of certain peripheral neurons. Finally, the fact that cutting the fiber connections with targets led to neuronal death added an entirely new dimension. It allowed Hendry and Iversen to picture a sequence of steps through which target NGF supports neuron survival.

Steps to Survival

The series of steps that they pictured actually constituted a new hypothesis. As so often happens in science, hypotheses multiply, as one creative idea leads to another and another. As knowledge advances, hypotheses propagate, and the new hypotheses lead to new knowledge. Hendry and Iversen's hypothetical sequence of steps begins with the production of NGF by target tissue. The NGF is then taken up at the ends (terminals) of the nerve fibers that have reached the target. Since cutting the long nerve fibers caused the distant nerve cell bodies to die, Hendry and Iversen proposed that the nerve fibers transported the NGF back to the cell body where critical, life-giving actions occurred. Because electrical impulses travel in the forward direction from nerve cell body to fiber, while NGF was pictured to travel backwards, from fiber to cell body, the process was termed "retrograde transport"(Figure 1-2). As a refinement of the NGF target hypothesis, the retrograde transport hypothesis tied together neurons, targets, NGF, and the growth and function of fiber pathways and circuits, hallmarks of the nervous system. The hypothesis was satisfying in its synthesis of diverse elements. Now the hard work of experimental verification, or falsification lay ahead.

Hendry was coming to the end of his doctoral project, but there was time for one or two crucial last experiments. The question was how to show that NGF actually is transported from the target, backward via the connecting fiber to the nerve cell body. This was particularly daunting, because it was still difficult to measure NGF reliably in the body. Hendry and Iversen used a time-honored alternative to directly measuring NGF. The trick was to label NGF molecules with a tag that can be easily measured. Then, the tag can be followed in the body to trace the travels of NGF. With the advent of radioactivity, the conventional approach involved tagging molecules of interest with radioactive atoms, and then simply following the radioactivity in the body, to trace the molecule of interest. If done properly, the body did not distinguish between a

Figure 1-2   The retrograde transport of nerve growth factor (NGF). In this model, NGF is synthesized by the target, taken up by the innervating nerve terminal, and transported back through the nerve process to the cell body where biological actions occur. Note that electrical impulses are propagated in the forward direction, from cell body to target terminals, while NGF travels in the opposite, retrograde direction.

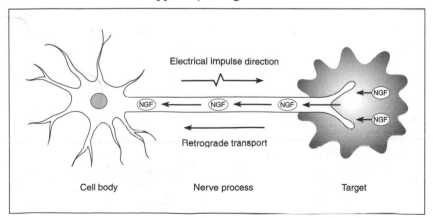

tagged molecule and its naturally-occurring counterpart and thus no distortion was created by the tagging.

Hendry's general strategy was to introduce the radioactive NGF into a target tissue and then determine whether the nerve fibers investing the target became radioactive. If they did, the next step would be to see whether the distant nerve cell bodies eventually became radioactive, presumably by means of retrograde transport through the connecting fibers. Hendry chose the eye as a sample target, since the iris is jam-packed with nerve fibers that were known to be dependent on NGF. The eye has the virtue of being easily accessible for treatment. Moreover, the nerve cell bodies that send fibers to the iris lie in the neck, at some distance from the eye. So there was little danger of inadvertently contaminating the neck neurons with radioactivity. Hendry prepared the sample NGF in a test tube, attaching the radioactive atom to tag the molecule. He then treated one eye of each rat with the tagged NGF, and used the opposite eye and nerves on the other side of the rat as controls. He waited a number of days to allow the iris nerves to take up the NGF and to allow the fibers to transport the tagged NGF back to the neck cell bodies. The results were again dramatic. The nerve cell bodies in the neck were filled with radioactivity only on the side where the eye had been treated with radioactive NGF. Nerves on the other side of the neck were not radioactive, indicating that the tagged NGF did not simply enter the general circulation after eye treatment, labeling body tissues willy-nilly. The experiment provided the first evidence that nerve fibers actually did transport the survival factor from target to cell bodies.

Many more experiments had to be done to validate these initial impressions, but Hendry was at the end of his doctoral studies. He decided to do a postdoctoral fellowship in Switzerland, where the pubs were allegedly substandard, but where the skiing was decidedly better than in England. He was about to determine whether skiing and neuroscience mixed as well as beer and neuroscience.

Switzerland

Hendry arrived with skis in Basel, Switzerland, to work at the Biocenter, in the laboratory of Hans Thoenen, a towering figure in European neuroscience. Thoenen towered in many ways. A lean, sinewy 6 feet 5 inches, he radiated intensity and focus. His craggy face seemed to have emerged fully formed from the remote alpine valley where he grew up. He had already worked his way through several distinct careers. As a university student, medical student, and medical trainee, he was a member of the national Swiss mountain climbing team that was credited with numerous ascents in the Alps and Andes. After seeing patients as a geriatric clinician, he joined the pharmaceutical giant, Hoffmann-La Roche, and tore through discovery after discovery with such intensity, that management tapped him for the executive suite. In typical fashion, Thoenen refused to leave the laboratory. When management persisted, he simply left Roche, to start a new, academic career and continue his science. Hans Thoenen was not a man of moderate opinions.

The agreed-upon challenge was perfectly clear to Hendry and Thoenen: perform experiments to demonstrate beyond any doubt that NGF was transported backwards from target to neuron cell body. While agreeing on substance, their styles exhibited creative contrasts, charitably speaking. Thoenen worked from predawn to postmidnight, only because other parts of the day had not yet been identified. His coworkers had never seen him idle. Hendry threw himself into the arduous task of skiing in the Alps with equal dedication. Experiments were rather precisely designed to conform to the all-important ski-lift schedules. Skiing and science seemed to mix, and in a few short months the Hendry-Thoenen international neuro-ski team had begun amassing evidence to support the retrograde transport hypothesis.

While the experiments were actually quite varied, a general theme emerged. The strategy was exemplified by a simple approach. A thin suture thread was tied around the fibers connecting the nerve cell body and the target. If NGF was, indeed, being transported backwards from target to nerve cell body, NGF molecules should pile up on the target side of the suture dam. That is exactly what happened. Whether the dam consisted of suture thread, a drug that prevents all transport of substances in the fiber, a fiber crush, or even freezing, NGF always piled up on the target side, indicating that NGF was traveling from target to nerve cell body. The experiment was performed with radioactively tagged NGF, with NGF normally produced in the target, and with combinations of the two, and results were always the same. The dam also prevented NGF from accumulating in the nerve cell body, as might be expected. Exhaustive control experiments indicated that the dam itself prevented the transport. Even extensive runs on black diamond (expert) ski slopes at high altitude could not obscure the fact that retrograde transport from target to nerve cell body was a fact of life in the peripheral nervous system.

While the following years would yield many important details that explain the process of retrograde transport, discovery of the basic principle represented a critical advance. It tied together many aspects of life in the (peripheral) nervous system. It was a first step in understanding the nature of nerve connections, and how connections form. Though not appreciated at the time, these insights would ultimately form the basis of a trail map for brain synaptic connections and learning, memory and mind. Since the stakes for understanding the brain-mind connection would prove to be so high, it might be useful to summarize some of the more important insights.

The discoveries placed a molecule, NGF, at the center of the formation of nerve connections that process information. In the broadest sense, the work pointed to a molecule-mind connection, though this was hardly a topic of conversation at the time. A plausible process was defined, in which nerve targets produce a hormone, NGF, that attracts distant nerve fibers. The fibers grow to the appropriate target and form connections. This explains directional fiber growth, pathfinding, and how targets are selected. In sum, this provides one explanation for a central mystery of the nervous system: how specific and selective circuits form. The work also unified the apparently unrelated phenomena of selective survival of neurons and successful connection. Only those neurons that contact the proper target, and gain access to life-giving NGF, survive; neurons that fail to make proper contacts die through NGF deprivation. Life and death in the nervous system are intimately related to connections that generate the mind. By implication, derangements of NGF could unravel these relationships, resulting in neuron death, loss of connections and, conceivably, loss of mind.

As always, the manifold insights led to new questions that emphasized the unfathomable depths of our ignorance. How does NGF support neuronal survival? How does NGF actually make nerve cell bodies grow larger? How does NGF cause nerve cells to elaborate long fibers and direct them to targets? How can a single molecule, NGF, have so many different effects on a single neuron? Are these apparently distinct biological actions related on some deeper level that we just do not understand?

Finally, how does this scientific tale explain the tragedy of Alzheimer's disease, that was robbing Enoch Wallace and Laurie Grundfest of their humanity?

To approach the last question, we continue by inquiring how NGF actually acts on neurons to evoke a dizzying array of effects.

Enoch's legendary vigilance, organization, and preparedness begin to fail: for the first time, his always impeccable wardrobe lacks panache, he becomes tangential in the boardroom and at home, misplaces the newspaper, and forgets appointments. Incipient dementia may reflect cell death and deranged nerve connections. In the laboratory, NGF, which mysteriously supports neuron survival and connectivity, becomes the focus of intense experimentation. To understand the factor's actions a search for its receptor engages scientists in multiple fields, from neuroscience to cancer research.

2

THE SURVIVAL SIGNAL

❖

TUESDAY MORNING Enoch Wallace found his way to the bathroom without event. He showered, shaved, dressed, and rushed to meet his driver, who navigated through granite canyons to the Wall Street area. Racing from conference room to boardroom to conference room, the parties parried, thrusted, stormed, and reconciled through the human ritual of negotiation. By 10 P.M., the rough outlines of the corporate acquisition had taken shape, and the engagement broke for the day. The confusion of Monday morning was not even a distant memory, and Enoch was driven to his East Side apartment with the exhausted satisfaction that follows bloody skirmishes that precede victory.

The next morning at a strategy session in the boardroom, Enoch seemed distracted, and uncharacteristically rambled while outlining his tactics for the afternoon's session. Over the ensuing weeks, his neckties were not quite as straight, his suits not quite as immaculate, and his shoes not quite as bright as those of Enoch Wallace, Esq. At home, he misplaced *The Wall Street Journal,* he neglected to return the call of Jeb Blanch, his partner in the negotiation, and his wife had to remind him about the dinner reservations for Saturday night. Over cocktails at dinner, his usually focused conversation became tangential when discussing the Lincoln Center chorale. With the acquisition on his mind, and the hectic prospect of an upcoming merger, he referred to Kerry as Terry on two occasions. While no single error was alarming, Enoch's legendary alertness, vigilance, focus, and articulate discourse were not quite in evidence.

Since his days as an accomplished technical rock climber in college, Enoch took pride in his informal, but rigorous preparedness, his attention to the details of equipment. On any steep pitch, a frayed rope, a faulty carabiner, a weakened piton could doom the entire climbing party. His partners had learned that Enoch attended to the details that ensured survival on the rock face. Though never extravagant, his tools of the trade were always battle-ready and reliably lifesaving. This implicit dependability carried over to his everyday life. His dress was practical, functional, appropriate, and attractive in its economy. Winter coats kept him warm, rain gear ensured dryness, and hats protected him from the sun. His Wall Street garb was no less effective. Quietly, ties matched suits, which complemented shirts and shoes effortlessly. It was impossible not to notice, consequently, when his ensembles began to fragment. Though inapparent to strangers, ill-matched ties, shirts for other suits and belts askew drew the attention of close associates, and former climbing buddies. Enoch was relaxing, or changing.

Sally, too, noticed changes, though she was more prone to denial than anyone else: there were small things, maybe nonexistent things. On a hike at the Vineyard, Enoch wore his beloved Vasque boots, that had seen him through hundreds of miles of wilderness, and through two changes of Vibram soles. Every fall Enoch waterproofed the boots for the season. Yet, as they walked through the semi-swamp terrain, Sally noted that the water was not beading on the boot surface; they couldn't have been snow-proofed. "My God, I must be going crazy," Sally mused silently. But somewhere, deep beyond the reach of conscious thought, neither visible nor palpable, Sally dimly sensed an emerging, terrifying truth.

The other thing was the walking stick. Enoch was a master of the walking stick, the virtues of which he never stopped extolling to any hiking partner unfortunate enough to express the most indifferent of interests. In fact, a returned glance alone, while Enoch commented on the stick, was quite enough to doom the trusting friend to a sermon on the stick. Not only important for the balance, rhythm, and pace of the hike, but a savior of the lower back, a support for a resting pack, a probe for deep nooks and crannies, a post for a makeshift tent, an instrument for handling snakes, and Oh, so much more, was man's best friend, the walking stick. So, Sally was actually shocked when they stopped for a break, and Enoch set the stick against the tree with tufts of sullying grass adhering to the working end. For the first time in memory, Enoch had not used his spare bandana to wipe the stick clean.

It would have been impossible to determine, at this stage, whether these marginal symptoms reflected mild biochemical dysfunction of the neurons at the base of his brain, whether the connections to the cerebral cortex were temporarily disordered, or whether some of the key neurons had already died. The Alzheimer's disease was progressing.

❖

Receptor locks on the surface of cells throughout the body lie at the heart of intercellular communication. Nerves communicate with their cellular targets by releasing molecular signal keys that selectively interact with specific receptor locks, eliciting biochemical changes and biological responses in targets. Not surprisingly, receptors play critical roles in a variety of diseases and treatments. The neural regulation of cardiovascular function is illustrative.

Andre Mallard is a forty-eight-year-old French-American chef with a family history of hypertension and heart disease. His father died suddenly at forty-six with the crushing chest pain of a heart attack; at age fifty, his uncle, Claude developed shortness of breath after walking two blocks and his doctor diagnosed congestive heart failure, treating him with digitalis to good effect; his aunt Nicole experienced cardiac arrhythmias just after menopause, which required treatment. Andre felt like a cardiac catastrophe in waiting.

Virtually on cue, Andre experienced his first symptoms at forty-five. While working on the dill in his herb garden on a hot, bright July afternoon, his heart suddenly began racing. Though lasting only seconds, he had a sense of breathlessness and the feeling that his heart could pop out of his chest. There was no pain. It all happened so fast that Andre thought he must have imagined the experience, and continued loosening the soil. He forgot about the episode. However, the next evening at dinner, he experienced a similar event. His heart began racing, he was distinctly out of breath, and even had a sense of faintness, though he never lost consciousness. The whole affair lasted several seconds longer than the arrhythmia in the garden. There was no denying that this was real. Andre's wife, Marie, was alarmed, and insisted that he visit their internist the next day. He had actually not seen the doctor for over five years. The trip to the doctor was revealing, perhaps lifesaving in the long run.

Dr. Devereaux asked Andre about chest pain at rest, after garden work, upon climbing stairs. The happy answer was "no" on all counts.

"How many pillows do you sleep on?"

After a moment's thought, Andre answered "two at times," and realized that he did sometimes feel short of breath on one pillow. And yes, on rare occasions he had noted some swelling of his ankles, though he couldn't relate that to any particular activity.

"No," he never awoke at night out of breath.

A long series of questions was followed by the exam, which began with blood pressure measurement and an electrocardiogram, worked its way through chest thumping (percussion) and stethoscope (auscultation), and ended with an office chest x-ray and blood and urine tests.

The postexam conference was illuminating. Dr. Devereaux's haywire fringe of gold-gray hair glowed in the window sunlight as he explained the findings. First and foremost, Andre was walking through life with a dangerously high blood pressure of 180/110, well above the upper limits of normal of 140/90 ("one-forty over ninety" in medical parlance). Dr. Devereaux guessed that a family history of hypertension was a strong factor contributing to the pedigree of heart disease. In Andre's case, the hypertension was already deleteriously affecting his heart. The EKG (electrocardiogram) revealed cardiac enlargement, mild ballooning of the heart, which interfered with efficient pumping. The stretching probably gave rise to the skipped beats as the electrical activity of the heart reacted abnormally. Andre's skipped beats were of the potentially dangerous type that affected the largest chamber, the left ventricle. This kind of arrhythmia has been known to cause cardiac arrest. The x-ray confirmed the cardiac enlargement.

Andre's cardiac enlargement was also contributing to his episodes of shortness of breath through inefficient pumping, and to his ankle swelling as backed up blood forced fluid out of small vessels in the legs, and gravity caused it to collect in the lowest part of his body.

In Dr. Devereaux's opinion the culprit underlying Andre's cardiac abnormalities was his hypertension. The good, lifesaving news was that hypertension is treatable. The internist sketched the physiological and treatment scenarios. Though they would use a number of medicines, the core of the therapy would be a blockade of the cardiac receptors underlying the hypertension. The heart receives connections, or is innervated, from sympathetic nerves that are part of the "fight-or-flight" system. The heart also responds to circulating adrenal gland hormones. The sympathetics release norepinephrine and the adrenal releases the circulating, closely-related hormone, epinephrine. In an emergency, these two chemical signals increase heart rate, the force of cardiac contraction, the amount of blood pumped and generally prepare the system for the sudden activity of fight or flight. All of these actions are unleashed when the norepinephrine and epinephrine signal keys open the beta receptor locks on the surface of the heart. While this can be lifesaving under rare threatening circumstances, chronic, abnormal beta receptor stimulation is damaging, resulting in the disorders experienced by Andre and his family. Beta receptor blocking drugs prevent norepinephrine and epinephrine from activating the heart's receptors, and thus stop the process of hypertensive cardiac damage. Used in conjunction with several other drugs, the deadly spiral of events is aborted, and patients can live a normal life.

Within days of beginning treatment, Andre's skipped beats disappeared, his shortness of breath was a memory, and over the ensuing weeks no ankle swelling occurred. A return visit to the internist revealed a blood pressure of 135/85 and the EKG was returning to normal. Beta receptor blockade had saved Andre from the fate of his forbears.

Chemical signals throughout the body exert biological effects by activating their matching receptors, as illustrated for norepinephrine and beta receptors. It was critical to find the NGF receptor to understand how NGF works. To begin unraveling the mysteries of NGF action, scientists across the world initiated a search for the receptor that mediated life and death, and the formation of neural connections.

The Nerve Growth Factor Signal

The NGF experiments, beginning at about the same time and initially having nothing to do with brain and mind, ultimately led to the discovery of physical mechanisms underlying the formation and function of nerve connections. Decades in the future, they would point to the physical basis of mind. By 1980, the work had already integrated the seemingly unrelated phenomena of neuron size, number, biochemical function, fiber growth, and life and death with the formation of connections. The capstone holding this entire edifice together was a single molecule, NGF.

One of the central mysteries concerned how NGF, acting on a single neuron, could exert so many different effects. We throw a switch and a light goes on, but the electrical signal does not also frame the house, put up plasterboard, paint walls, lay floors, hang mirrors and pictures, install plumbing, flush toilets, and serve dinner. In the absence of NGF the whole neuron ceases to exist, but a house remains even if we don't throw the switch. The more we learned about NGF, the larger our collective scientific headache became.

What questions would bring us closer to the solution of this puzzle? One was to ask how NGF interacts with neurons. This question's significance was appreciated by Eric Shooter, a scientist who has taken part in innumerable advances in the understanding of NGF. Shooter, born and educated in Great Britain, emigrated to the States and established himself as a biochemist of the first rank. He was a dignified, articulate, classically understated Briton. His reserve did little to conceal a generosity of spirit, a warmth and a respect for colleagues that ennobled any group of which he was a member. By the time he retired from Stanford University in 1994, he had trained, guided, and nourished generations who carried the growth factor field from the fringes to the core of biology.

Shooter began his scientific career as a protein chemist, a classical biologist concerned with the molecular workhorses of life. Protein enzymes perform chemical transformations that are the substance of life; proteins are the building blocks of cells and tissues; most of the signals that cells use to communicate with each other are proteins. However, Shooter spent his graduate career studying proteins that were far more unusual, exotic, and perhaps less central to human life than those we have just described.

As a doctoral student at the University of Cambridge, the future growth factor pioneer and neuroscientific great was given a somewhat bizarre charge. He focused his formidable intellect on solving the wondrous mysteries of proteins of a unique species: the peanut. As Shooter relates, his mentors had visions of revolutionizing the world with peanut proteins. The pedestrian peanut contains proteins that form long, strong fibers in the purified state. Shooter's mentors had grand plans to use the peanut protein fibers to manufacture clothing. It would be the most important revolution since the spinning wheel. Suits, dresses, jackets, coats would be made of peanut protein. Wool and cotton would become obsolete. As Shooter labored at the laboratory bench purifying and characterizing the fibrous proteins, plans were being made to grow peanuts on an industrial production scale at British holdings in West Africa. While Shooter's scientific experiments progressed quite well, several unfortunate accidents of history derailed his professors' dreams of garment grandeur.

The fauna of West Africa turned out to be singularly uncooperative. While the peanuts grew, the insects and assorted vermin of the dark continent decimated the crops. Not a peanut was left at the end of the growing season. Natural disasters, however, could have been mastered by the hardy, resourceful British. The island nation had transcended innumerable calamities in its long history. The death knell for the peanut plan was a man-made tragedy that could not have been anticipated: Nylon. Just as plans were maturing for peanut suits, peanut hats, peanut dresses, and peanut pants, synthetic fibers were launched like missiles across the Atlantic. While Shooter's studies were successful, and earned him a coveted Cambridge doctorate, peanut protein haberdashery never saw the light of day. George Washington Carver's place in history remained secure. After a number of visits to the States, Shooter settled at Stanford, and began his pioneering work in an area that was, at the time, only slightly less exotic than that of peanut proteins. In the 1960s, growth factors were only marginally less risky than peanuts.

Search for the NGF Receptor

Now an accomplished protein chemist, Shooter turned his attention to the NGF receptor, reasoning that any hope of understanding how NGF evokes its multifarious effects on neurons must begin by grasping how an NGF protein molecule interacts with a neuron. When he began his experiments, it was generally believed that signals such as NGF acted as keys that fit receptor locks on cell surfaces. These keys and locks initiate biochemical events that evoke responses from the cell. At this time no one had isolated and characterized a receptor, though overwhelming indirect evidence indicated that they must exist.

To probe for the receptors, Shooter radioactively tagged NGF and measured its binding to the external surface cell membrane of neurons. The surface membrane encloses a cell, facing its outside world and receiving incoming signals (Figure 2-1). Shooter's strategy was to mix infinitesimal amounts of NGF with these membranes and measure the NGF that bound strongly to them. With control experiments, Shooter could identify the binding attributable to receptors. He could surmise receptors' characteristics from the concentrations of NGF necessary for binding, and he could determine the nature of the binding.

Painstaking experiments ensued. The results provided the first clear glimpse of the nature of the NGF receptors. Shooter's team noted that NGF-binding to the surface membranes was particularly strong at two

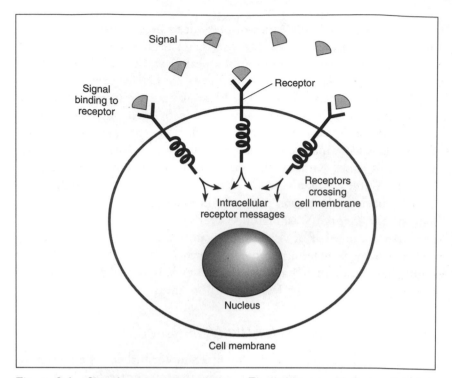

Figure 2-1   Signal-receptor interactions. The nerve growth factor signals elicit actions by binding to specific receptors projecting through the surface of the cell membrane. Signal binding activates receptors, generating intracellular messages that evoke biological responses.

concentrations of NGF.[1] Binding occurred at low concentrations—one part NGF in 100 billion. This implied a receptor with a high affinity for NGF; it bound NGF extremely tightly and was dubbed the high-affinity receptor. Binding also occurred at a hundredfold higher concentration of NGF—one part in a billion—and indicated the presence of a low-affinity receptor. From Shooter's indirect approach, two different receptors were inferred, though they had not been isolated or identified. It was like deducing the presence of a fox from tracks in the snow, or a person from a shadow on the cave wall.

Which receptor was responsible for which biological effects of NGF? This was the big question in trying to unravel the multiple actions of NGF on neurons. Fortunately, there was a simple way to answer the question. If an action occurred at low NGF concentrations, one part in 100 billion, it must be attributable to the high-affinity receptor. At such a low concentration, the low-affinity receptor would not even bind NGF. Yet, if high concentrations of NGF were required for an action, the low-affinity receptor was probably involved. The hunt began.

Life and death represented the most profound sphere of actions of NGF on responsive neurons. Experiments revealed that NGF was an absolute requirement for the survival of many groups of peripheral neurons lying outside of the brain. Now the question was, which receptor mediated the life-giving actions of NGF? The new experimental results were unequivocal. NGF at one part in 100 billion rescued neurons from death. Shooter's experiment was repeated over and over in tissue cultures across the globe. Binding the NGF key to the high-affinity receptor lock supported the life of neurons; that is, astonishingly low concentrations of NGF gave life. Action at these vanishingly low concentrations could hardly be a nourishment effect. Somehow stimulation of the high-affinity receptor must have approached the secret of life.

The next question to answer was, which receptor mediated nerve fiber outgrowth, the first step in forming connections? The answer, essential to understanding the molecular basis of neural circuitry, could provide crucial insights into the generation of brain architecture, brain function, and mind architecture. The experiments were again performed in tissue culture to ensure rigorous control. Neurons in culture dishes were exposed to different concentrations of NGF, and nerve fiber growth was measured under the microscope. The results were crystal clear: The low concentration of one part NGF in 100 billion was enough to generate maximal fiber growth. The interpretation was equally clear: Binding NGF to the high-affinity receptor made the nerve fiber grow.

The impact of this discovery is difficult to overestimate. For the first time, a specific receptor, presumably a specific molecule, could be associated with the central fact of the nervous system: connectivity. Although the receptor had yet to be isolated and characterized, physical structure and brain function might be bridged, at least in principle. This experimental finding held out the promise that an unbroken line of logic might link molecules, neural connections, nerve circuits, brain architecture, and mental function. The details of the experiments were equally revelatory. NGF evoked nerve fiber growth at vanishingly low concentrations. One molecule in a sea of 100 billion other molecules (one in $100,000,000,000$ or 10^{11}) was enough to generate the connections that underlie consciousness.

As each action of NGF was studied, the high-affinity receptor held the answer. The NGF-associated changes in neuron size, shape, number, biochemistry, and electrical properties, for example, all occurred with low concentrations of the growth factor. Hence the high-affinity receptor became known as the biologically active receptor, and the low-affinity receptor remained in the shadows as a mystery. It was still a mystery how the single biologically active receptor could perform all these actions at the same time. Also, how could fleetingly low concentrations of NGF exert such powerful actions, from life and death to creating connections? The scientific progress raised more questions than answers, but, as always, the questions guided future experiments. It was going to be necessary to isolate and examine the receptors. Indirect methods of measuring NGF-binding to neuronal membranes and making complex calculations would take us only so far. Ever finer measurements and calculations lead only to ever finer speculation in a world of virtual receptors.

The task of isolating the NGF receptors from the thousands of other proteins in neurons was daunting, but we did have a few clues as to the nature of the receptors that suggested how to proceed. Since NGF bound tightly, it was possible to use radioactive NGF as a label to follow the receptor-NGF link through biochemical procedures. Improving on nature, we could use chemicals to render the receptor-NGF link permanent. Then it was easy to follow the radioactive label, forever linked to the receptor, through chemical reactions in the test tube. This is a convenient way to learn valuable things like a receptor molecule's approximate size—a critical piece of information because biochemical techniques can separate, purify, and isolate molecules by size. We could then determine whether a molecule qualified as the receptor simply by determining if it was the right size. Similar approaches provided indirect information about shape, electrical charges, and how the receptor binds the NGF. These fragmentary insights, along with the well-known biological actions of the receptors, would help to isolate the receptors—if only a powerful enough approach could be devised.

Receptor Search Continued: A New Approach

Molecular biology provided the tools. In the 1980s and 1990s, genetic engineering was sweeping through neuroscience like a tidal wave. With breathtaking rapidity, genes were being cloned for Huntington's disease, muscular dystrophy, syndromes of mental retardation, and forms of Alzheimer's disease. New molecular approaches were revealing the diverse genetic identities and functions of billions of neurons in the brain. In one notable scientific campaign, genes for the acetylcholine receptor were cloned. This receptor, the lock for the acetylcholine neurotransmitter key, makes muscles contract and thus causes all voluntary movement. Defects in the acetylcholine receptor result in diseases of paralysis, such as myasthenia. The successful cloning of this receptor provided a roadmap, a strategy for cloning other receptors. The stage was set for a molecular attack on the NGF receptor. But a stage is accessible only to those with the intuition and insight to recognize the scenery. Often what is needed is the fresh vision of a newcomer, an outsider who is not blinded by the familiar. Moses Chao was such a neuroscientific outsider.

Chao grew up in the new area of molecular biology, far from the conundrums of brain function. After graduating from college in California, he entered graduate school in chemistry at UCLA, a sprawling campus regarded as a paradigm of the American vision, forever building and expanding. As a graduate student, he studied the intricacies of gene structure and function. His work explored how DNA and proteins are put together deep in the nucleus within the cell. He showed that codes of DNA directed binding to special proteins, which, in turn, governs gene action. He studied repressors that turn off gene action and factors that regulate repressors in microstructures within the nucleus. Although UCLA is home to the huge, world-class Brain Research Institute, Chao dwelled in the realm of the nucleus, light years from the brain.

He was accepted to the illustrious laboratory of Richard Axel at Columbia College of Physicians and Surgeons in New York for postdoctoral study. There he was exposed to people and ideas that ultimately seduced him into the molecular biology of the nervous system. Axel, a precocious, brilliant molecular biologist, often joked about his medical student days at Johns Hopkins. With self-deprecation, he noted that his teachers, after careful observation, encouraged him to maintain distance from patients, to minimize the damage he might do. His advisors, according to his stories, guided him to the specialty of pathology, where havoc could be avoided by working with dead patients only. After watching him in the morgue, though, his mentors rethought their advice and concluded that he was quite capable of even damaging the dead. They recommended that he maintain a low profile in the laboratory. This comical account of his medical odyssey could not obscure the brilliance of a rising scientific star.

Drawing on his graduate experiences with gene structure and function, Chao investigated how genes turn on and off in mammalian cells. He concentrated especially on the part of the gene that controls its level of activity, the so-called promoter. When the promoter is stimulated, the gene turns on and produces its product. When the promoter is inhibited, the gene turns off. Chao chose the globin gene as a model system. Globin, a protein produced by this gene, forms part of the hemoglobin molecule that carries oxygen to tissues throughout the body. Though the gene is significant in its own right, it also introduced Chao to major new concepts and approaches.

To study how the gene turns on and off, Chao introduced foreign globin genes into leukemia cells. The experimental technique, known as gene transfer, enables an investigator to follow the activity of a foreign, easily identified gene in a cell with its own globin genes. This useful technique would eventually enable Chao to isolate the NGF receptor.

In addition to its advanced techniques, the Axel laboratory was also endowed with creative people and an innovative atmosphere. During Chao's tenure, the laboratory had initiated a new collaboration with a neighboring neuroscience center, Columbia's world-famous Center for Neurobiology and Behavior. Its director, Eric Kandel, and his collaborators had pioneered the use of simple animal systems to study mechanisms of learning and memory. Kandel discovered that even the lowly sea snail, *Aplysia californica*, was capable of simple forms of learning. He concentrated on the snail's simple nervous system to discover basic cellular mechanisms of memory. Though rather deficient in Shakespeare's sonnets, world history, and higher mathematics, the snail could associate a stimulus with reward or punishment, react accordingly, and remember.

Kandel, ever at the cutting edge, collaborated with the Axel group to approach the molecular biology of the snail nervous system and, ultimately, of learning and memory. The first step was to define the neurotransmitter signals that snail neurons use to learn and remember. The scientific work at the laboratory bench was performed by a postdoctoral wizard from California, Richard Scheller, no ordinary, green postdoc. He was one of the original scientists at the high-flying California biotech company, Genentech. After notable successes at Genentech, he longed for the academic life and entered Axel's lab. In breathtaking fashion, Scheller was the first to clone genes for snail neural signals that shed light on the molecular basis of memory in aplysia—work that turned the attention of Axel and his coworkers to the nervous system. Chao began to think about applying molecular approaches to the nervous system.

As he came to the end of his tenure in the Axel laboratory, it was time for Chao to choose an independent scientific problem. This critical juncture in a scientist's career requires taking a careful inventory of internal abilities and external possibilities. Luck doesn't hurt either. Chao enjoyed a happy coincidence of interests, experiences, talents, and intuitions. He had worked with tumor cells and was well aware of the exciting unknowns surrounding growth factors and tumor growth. He was expert at working with genes, and with DNA in general, and comfortable using gene transfer techniques to study gene action. He was convinced that neuroscience was the wave of the future. His intuition led him to growth factors and the nervous system. That path led unerringly to the nervous system's only factor of the moment, NGF—the answer to the question of his scientific direction. The obvious follow-up question was how to find the NGF receptor. With that settled, there only remained the small detail of solving the problem.

Chao joined the faculty at Cornell University Medical College on the Upper East Side of Manhattan and began plotting his strategy in this citified bucolic setting. There, he came up with the novel idea of taking a gene transfer approach to isolate and clone the NGF receptor. In essence, he would extract DNA from human cells known to bind NGF and transfer the DNA to animal cells devoid of binding. The host cells with the foreign DNA should make the receptor, which could be identified by NGF-binding and by use of antibodies that only recognize the human receptor in the animal cells. Then the NGF gene could be isolated from the unique, foreign DNA in the host cell.

Armed with this imaginative approach, Chao took the next step in a fledgling scientist's career. He applied for grant money to support the work. The newly minted independent scientist faces a potentially lethal Catch-22 when applying for his first grant. He has no independent track record of scientific achievements to assure the review committee of scientists that he can accomplish the work. But, without a grant, the work cannot be performed, and the track record cannot be established. An additional paradox greets even seasoned scientists who apply for funding. If a proposal is too novel, too unorthodox, it may be deemed far-fetched and unachievable. Yet standard approaches are often considered dull and not worthy of funding. All in all, it is far easier to criticize and kill an idea than to go out on a limb, take a chance, and support the unorthodox.

The evaluating committees initially thought that Chao's ideas were too risky, especially for an unproved scientist. Nevertheless, he managed to scrape together enough resources to perform the critical first experiments.

Chao's experimental strategy required special types of cells and exotic chemicals, all of which were individually available. Two human tumor cell lines, neuroblastomas and melanomas, were sources of NGF receptor genes; these tumors were known to produce the receptors. These are virulent forms of cancer. Neuroblastomas commonly arise in the eyes of infants and children, and frequently lead to blindness. Melanomas derive from close relatives of neurons, melanocyte pigment cells, usually of the skin and are highly malignant. Human cells were chosen because monoclonal antibodies to the human NGF receptor were available. Antibodies recognize the human NGF receptor but not receptors from other species: the antibodies can detect the production of human receptors in cells from other species after gene transfer. Chao took DNA from human cells and transferred it into mouse cells, which did not produce any NGF receptors of their own. The mouse cells containing human DNA were then grown in culture dishes.

Chao and his team had two ways of testing for the production of human receptors in the mouse cells. First, they could use the antibodies. Second, they could examine NGF-binding. One approach could confirm the other. The experimental procedures were tricky, and could go wrong in many ways. The DNA had to be carefully removed from the human cells to avoid damage; the DNA had to be transferred to mouse cells intact; the mouse cells had to be grown in just the right medium for the right amount of time at just the right temperature to prompt the growth and production of receptors; the receptor detection procedures had to be precise, finding receptors but not mistaking other molecules for receptors.

It all worked! The antibody detected the human NGF receptor in the mouse cells. The radioactive NGF bound to the presumed receptors on the mouse cells. The human NGF receptor gene had been successfully transferred to the mouse cells. The receptor and its encoding gene could now be analyzed to understand how NGF exerts its dramatic actions in the nervous system. The prospect of gaining insight to mechanisms of life and death and the formation of synaptic connections was intoxicating. They had reached a crucial juncture in understanding brain architecture and mind architecture.

Using DNA coding sequences unique to human DNA, the team performed experiments that isolated the NGF receptor gene. The human DNA containing the receptor gene was first isolated from the host cell mouse DNA, and then the human receptor gene was isolated. This gene was transferred to fresh mouse cells so the receptor could be analyzed in relatively pure form. The first step in characterizing the receptor, and its mysteries of life, death, and synaptic connectivity, was to define its interaction with NGF itself.

The team eagerly mixed radioactive NGF with the receptor-containing cells. Experiments were performed in several ways to avoid any possible error. The data were collected, subjected to calculation, and analyzed. The results were disconcertingly definitive. The isolated receptor only bound NGF at a relatively high concentration of one part in one billion. The scientific team had isolated the low-affinity receptor that had no known biological activity. Seven scientists from three different institutions had been working around the clock and now were the proud possessors of a receptor with . . . what? Summoning up optimism from the communal scientific well, they concluded their scientific report by speculating that some other form of this receptor might exhibit the desired biological effects. At least they had used a new approach to isolate receptors.[2]

As so often happens in science, Chao's team was in good company. The ubiquitous Eric Shooter and his colleagues were on the same trail. In fact, they also used a gene transfer approach to isolate and clone the same low-affinity NGF receptor. Within months of each other in the winter of 1986–1987, the Chao and Shooter teams each deciphered the structure of the NGF receptor gene. Although their results agreed and their structures were similar, their discoveries only added to the growing puzzle of how NGF worked.

Biologists are so keenly interested in structure, the code of genes, because it determines function. Structures of newly discovered genes are compared to genes of known function to deduce the function of the new genes. The NGF receptor gene, unfortunately, did not resemble any other known receptor genes, so Chao and Shooter's NGF gene offered no hint as to how NGF worked.

Receptor molecules generally stretch across a cell's outer membrane (see Figure 2-1). The outer portion of the receptor faces the world outside the cell and binds signals such as NGF. The inner portion of the receptor usually contains a relaying mechanism that changes the cell's biochemistry only when the outside signal (NGF) is bound (see

Figure 2-1). In this case, we would expect the NGF to bind to the receptor and stimulate its inner portion to send signals inside the cell because it was already known that receptors contain inner structures that enable them to do this. But the NGF receptor contained none of these specialized structures. In spite of all the work and imaginative speculation, cloning the low-affinity receptor elucidated nothing about how NGF exerts its remarkable effects. The NGF key fit the receptor lock, but there did not seem to be any opening mechanism.

The deepening mystery had no discernible dampening effect on the scientists' enthusiasm, however. They proudly christened the receptor with a new name that endowed it with an unwarranted air of understanding. Since it was a protein with a molecular size of 75 (kilodaltons, a measure of mass), it was renamed p75. Somehow, p75 sounded far better than the tacitly disparaging term "low-affinity," which implied rejection or, at least, mild undesirability.

The availability of the gene and the protein allowed laboratories throughout the world to measure the p75 receptor in every conceivable tissue, under every imaginable circumstance. This wealth of information, of limited use at the time, would lead to key insights in the future. However, while the data proliferated, the central mystery of NGF action remained.

The protein p75 was an actor in search of a role. A part in the grand NGF script of life and death, nerve fiber growth, and neural connectivity seemed totally out of the question. The character was all wrong. p75 bound NGF with a hundredfold lower affinity than required, and it possessed no known mechanism for translating the NGF embrace into any significant biological effect. As it turned out, the real play was unexpectedly going on in a distant theater.

A Little Help from Cancer Biology

As our understanding of the young science of biology deepens, relations emerge among disconnected fields. The big break in the NGF receptor impasse emerged from arcane work in cancer biology. In the early 1980s, cancer scientists had no idea that their work was destined to revolutionize our understanding of the nervous system.

One of the seminal triumphs in cancer research was the discovery of genes that cause cancer, the oncogenes. Though they remained elusive through long decades of searching, oncogenes had finally been cloned and characterized, and were the subject of a well-deserved Nobel Prize. The discovery of oncogenes, many of which caused unbridled cell proliferation as growth factors, led to a whole new field of research focusing on how these tumor genes work. Oncogenes seemed to participate in cancer arising in organs throughout the body. No single tissue or organ had a monopoly on oncogenes. Rather, the uncontrolled action of an oncogene in virtually any tissue could give rise to cancer of that tissue. So researchers studied tissues from lung to lymph nodes to liver, defining the roles of oncogenes in the genesis of diverse cancers.

New oncogenes were discovered by serendipity—and systematic screening. One common cancer, colon cancer, affects the large intestine, and the story of its discovery in 1986 illustrates how nearly routine experiments in one field rapidly revolutionize another. Scientists at the National Cancer Institute in Maryland were screening for oncogenes by transferring genes to a test group of cells; they wanted to determine which genes transformed the cells into cancer. They obtained a biopsy from a patient suffering from colon cancer, isolated genes, and identified one that made the test cells become cancerous.

The colon cancer gene was especially interesting. Although it did not closely resemble any other oncogene that had been identified, it exhibited many intriguing traits. Its structure resembled a receptor gene's; it coded for a segment facing the world outside the cell, a membrane-spanning segment, and a portion inside the cell designed to send intracellular signals (see Figure 2-1). The intracellular, signal-sending segment was of special interest because it was a member of an enzyme family that sends signals by adding phosphate molecules to proteins. This family, the kinases, rang biological bells because kinase receptors were known to be locks into which growth factor signal keys fit. For example, insulin, of diabetes fame, acts as a growth factor by binding to kinase receptors. Under normal circumstances, a kinase receptor would send signals inside the cell only when it was stimulated by an extracellular signal. (The identity of these signals was unknown in 1986.) In a freak occurrence of nature, segments of another gene, coding for a protein called tropomyosin, had inserted themselves into the middle of the kinase receptor gene. This abnormal kinase receptor gene was dubbed "tropomyosin receptor kinase," or trk for short (pronounced "track"). Because they contained these tropomyosin segments trk genes potentially override the mechanisms that keep kinase receptors normally inactive without outside signals. Thus in its abnormal state the trk gene keeps the kinase enzyme in a perpetually active state, leading to uncontrolled cell division and cancer. Kinases were known to cause cell division.

Colon cancer occurred when the tropomyosin gene segments were inserted into the normal trk gene present in all of us. Upon insertion of abnormal segments, the normal cellular trk "proto-oncogene" converts into the lethal oncogene. The abnormal segments turn on uncontrolled kinase activity and lead to colon cancer. But what was the function of the trk proto-oncogene in the first place? How can we figure out the normal function of a new gene that has little resemblance to the familiar?

One clue to a gene's function can be discerned from knowing where in the body the gene is active. As soon as trk was discovered, tissues throughout the body were examined to detect trk activity. Activity was undetectable in most tissues. However, trk activity was detected in the embryo, in developing sensory neurons. These neurons were the ones that Levi-Montalcini had discovered were sensitive to NGF. The significance of the findings was hardly lost on the cancer researchers. They hypothesized that the trk receptor might be the long sought after NGF receptor.

This strange idea from cancer biology gained credence in 1992. Levi-Montalcini had found that sympathetic neurons, the fight-or-flight system, also depended on NGF for survival, nerve fiber outgrowth, and the formation of connections. If trk was really the biologically active NGF receptor, it should be detectable in sympathetic neurons. Mark Bothwell at the University of Washington, a former student of Eric Shooter, found this to be so, which infused the trk engine with greater momentum. But the experiments were simply descriptive, indicating that trk was in the right places at the right times. It had not yet been shown that trk interacted with NGF and created a panoply of biological effects.

It was time to gather the battalions and mount a direct attack on trk. Chao and colleagues joined forces with Cancer Institute scientists. They employed a multipronged approach that relied heavily on cultured cell lines. In fact, their task may have been all but impossible in the absence of a particular, specially designed tumor cell line.

On the Importance of New Tools in Science: A Brief Digression

In science, as in all other endeavors, progress is often dependent on recognizing an unmet need, on discerning that a problem exists, on identifying an obstacle that was formerly invisible. And solutions to the newly appreciated problem are often revolutionary in impact and long-lived, whether complicated or simple in conception or design. Can we even picture the course of agriculture without the plow, the sledge, and the wheelbarrow? Can we conceive of Asian history and the Mongolian invasions in the absence of the lowly stirrup? Well, the NGF field had a huge black hole of a problem at its core. Peripheral sympathetic and sensory neurons were used in culture to work out the details of NGF action. However, these neurons are absolutely dependent on the presence of NGF for survival. So, NGF must be present in the medium to allow the neurons under study to survive, which made it impossible to grow cells without NGF to act as controls. With NGF present at all times, it was difficult to see how its various effects could be teased apart. Any observed action in the cells—increase in cell size, elaboration of fibers, alteration in cellular biochemistry—elicited by NGF might just be different faces of enhanced survival. Since NGF had to be present to allow the experiment to be performed in the first place, how could actions independent of survival ever be definitely identified? It was the genius of Lloyd Greene to recognize this central problem and devise a solution.

Greene had obtained his doctorate at the University of California, San Diego, with Sylvio Varon, a long-term collaborator of Levi-Montalcini. Varon was well versed in the tissue culture cult. Greene's training prepared him for the ordeals of NGF in the culture context. He moved east for his postdoctoral training, and joined the laboratory of Marshall Nirenberg at the National Institutes of Health (NIH) in Bethesda, Maryland. Nirenberg had just unraveled the DNA code, for which he would win the Nobel Prize. Like so many other molecular biologists, Nirenberg was in the process of leaping to the next frontier, neuroscience. While in Nirenberg's lab, Greene brought together the basics of neuroscience and tissue culture. He also became a friend of another postdoctoral trainee, Al Gilman. Gilman was intrigued by the general problem of how a receptor on the outside of a cell communicates with the interior. His solution to this problem was to earn another Nobel Prize several years in the future. Greene left the NIH for a Harvard Medical School assistant professor position, armed with his neuroscientific skills and curiosity about receptor mechanisms.

Greene wanted to study NGF and the way it communicates with neurons, but like many others in the field, he was confronted with the old experimental black hole: How can the many actions of NGF be understood, if the test neurons depend on NGF for survival? How can the effects on neuron growth, fiber outgrowth, and changed biochemistry be understood as individual responses, for example, if NGF always had to be present for the very same neurons simply to survive? Greene decided: he needed a nerve cell that could survive independent of NGF, but that still responded to added growth factor characteristically. Easier said than done. As he thought about the problem further, one potential solution recurred. Could he find a tumor cell, related to NGF-responsive neurons, that grew on its own, as cancers do, but still responds to NGF? Posing the question in this fruitful manner began the search that was to result in a quantum leap in neuroscience.

The quest was not a simple one. Cancer of neurons is actually exceedingly rare, perhaps because neurons do not ordinarily divide and reproduce, rendering them less susceptible to unbridled growth. Cancer of peripheral sympathetic or sensory neurons, the classic NGF targets, is virtually unknown. Greene collaborated with a neuropathologist at Beth Israel Hospital in Boston, Arthur Tischler, who was familiar with cancers of the nervous system. Together, they focused on a cancer of the adrenal medulla. Cells of the medulla are close relatives of sympathetic neurons, secreting the hormone epinephrine (adrenaline), a sib of the sympathetic norepinephrine transmitter. They obtained a specific adrenal tumor, known as a *pheochromocytoma*, or PC, for short. Adrenal medullary cells were known to respond to NGF by extending fibers in culture. Greene and Tischler reasoned that some of the cancer cells might retain their ability to respond to NGF, yet grow independently of the factor due to their cancerous nature.

They obtained a particular PC from colleagues at the New England Deaconess Hospital, that had been growing ("carried" for research purposes) under the skin of rats. Greene and Tischler placed individual PC cells in culture and isolated thirty-eight individual cellular clones as colonies. From these, they grew one special clone for seventy generations, which they designated "PC12." Colonies derived from the clone grew, divided, and multiplied in the *absence* of NGF, fulfilling the first criterion. When they added NGF to the medium the cells stopped dividing. Not only that, but they extended fibers just like normal neurons. After several weeks of NGF treatment, the fibers were 10 to 100 times the length of the cells themselves. Withdrawal of NGF resulted in degeneration of the fibers, beginning within twenty-four hours, just like normal neurons. However, unlike normal neurons, the PC12 cells also began dividing again, indicating that the effects of NGF were reversible. The cells also exhibited many of the biochemical responses to NGF that had been defined in normal neurons. Greene and Tischler had engineered a remarkable triumph.[3]

The general promise at the conclusion of their scientific article has been amply fulfilled:

> The PC12 line appears to be a useful model system for the study of numerous problems in neurobiology and neurochemistry. These may include the mechanisms of action of NGF and its role in development and differentiation of neural stem cells; initiation and regulation of neurite (*nerve fiber*) outgrowth; and metabolism, storage, uptake and release of catecholamines.[3]

Return to the Receptor Search

Chao and the Cancer Institute team relied heavily on the PC12 cells in their trk attack. With PC12, investigators had already uncovered a potential trk link. One of the earliest responses of PC12 cells to NGF-binding was the activation of kinase enzyme activity. Early experimenters suggested that trk kinase itself might be activated in PC12 cells by NGF. Preliminary impressions even indicated that appropriate concentrations of NGF did the trick.

The next step was to examine the PC12-trk-NGF response in greater detail. NGF-binding prompts the kinase enzyme part of the trk receptor to add phosphate molecules to itself in a process known as autophosphorylation, which in turn activates the trk and enables it to send signals inside the neuron. Indeed, adding NGF to PC12 cultures rapidly caused trk autophosphorylation.

A mutant PC12 cell line, which had lost its high-affinity biological responses to NGF, was also examined. These mutant cells no longer elaborated neural processes in response to NGF, and they could not be rescued by the factor. On examination, these mutants had barely detectable levels of trk. In a critical experiment, NGF did not elicit detectable phosphorylation of the extremely low levels of trk in these cells. The key experiment was then performed. Active trk genes were transferred to the mutant cells to determine whether production of the candidate NGF receptor could rescue the mutants. The results were dramatic. Synthesis of the introduced trk restored the ability of the mutants to respond to NGF. The cells now grew long neural fibers in response to added factor. NGF increased cell size after the factor was added. The most striking aspect was that the transfer of the foreign trk receptor gene allowed the mutants to survive in the presence of NGF. Thus, at least in this tumor cell line, trk and its autophosphorylation appeared to be necessary for high-affinity biological responses, including life and death.[3]

Could NGF also induce trk autophosphorylation in other trk-containing neuronal cell lines? Two different neuroblastoma lines were used, and NGF elicited trk autophosphorylation in both. So the trk response to NGF appeared to be generalized, and not restricted to a particular cell type.

The experiments next moved from the cancer line cell models to normal neurons. Sensory neurons that Levi-Montalcini had found were dependent on NGF for growth and survival, were examined. The sensory neurons were removed from mouse embryos and grown in culture. Addition of NGF resulted in nearly immediate autophosphorylation of trk.[4] In sum, the trk receptor in normal neurons and in their cancer counterparts responded to NGF; the evidence was mounting.

An extreme test of the trk receptor hypothesis was next performed. Could the trk molecule act as the NGF receptor in a totally alien cellular environment? As a test, an active form of the trk gene was transplanted into cells that are not neurons and do not normally produce trk receptors. Skin fibroblasts were used. In normal circumstances, fibroblasts do not respond to NGF. But the trk transplanted to the fibroblasts also responded to NGF with autophosphorylation. This proved that no other neuron-specific substances were necessary for NGF actions on trk. The trk acted like the NGF receptor even when isolated from its normal environment.[4]

In a final series of experiments, radioactive NGF bound directly to trk. The NGF-trk complex was isolated. The size of the trk molecule was determined by biochemical methods. It conformed nearly perfectly to the size of the biologically active, high-affinity receptor that Shooter had identified years before. The puzzle was almost complete—almost, but not quite.

It remained to be demonstrated that trk bound NGF at the high-affinity concentration of one part in 100 billion. This it did not do. Careful binding studies revealed that trk bound NGF at the same low-affinity concentrations as did p75. So immense progress had been made; the biologically active NGF receptor had been identified, cloned, sequenced, and manipulated. The receptor mediated life and death and the formation of connections among peripheral neurons outside of the brain. Yet the account was still incomplete. Was some combination of trk and p75 necessary for high-affinity binding? Was some other form of trk required? Was some other crucial element missing? Regardless, the biggest mystery of finding the key and lock mechanism that mediated survival, formation of connections, and, presumably, important aspects of neural architecture, had begun to yield to understanding.

Questions of Receptor Function

The identification of the receptor orchestrating neuronal survival, fiber growth, connectivity, and architecture, forced a central question into bold relief. What do these different processes have to do with each other? Why did evolution, in its wisdom, link these unrelated events? In molecular terms, how does the single trk receptor serve life and death, connectivity, circuit formation, and neural communication at the same time?

These questions are of pressing relevance to the relations of brain and mind, and to their unraveling in disease. Is Alzheimer's primarily a disorder of fiber growth, of connections themselves, of communication itself? Or is neuron death independent of these processes? Is Alzheimer's attributable to abnormalities of survival factors such as NGF, or receptors such as trk? Effective treatments depend on answers. Enoch Wallace's fate depends on getting the relations right.

Scattered clues suggested how trk might support survival, connectivity, and architecture at the same time. The first hint was indirect and followed from an experiment we already described. The active trk gene had been transferred to fibroblast skin cells, which usually do not make the receptors; naturally, these trk-less cells do not normally respond to NGF. The transplanted trk gene made the receptor even in these foreign cells. The trk receptor even responded to NGF in the alien environment with autophosphorylation. So far, so good. The trk is functional by itself in the absence of the normal neuronal milieu. But the fibroblast cellular response to trk was anomalous. In the foreign fibroblasts, trk stimulation caused cell division, not the survival response characteristic of neurons. Cell context appears to determine cell responses to the identical, functioning trk receptor. So the trk receptor does not cause neuronal survival, fiber outgrowth, and connectivity all by itself. It must get help from biological friends. There must be other molecular players that are cell-specific.

These new experimental observations accorded with already available information about related receptors. The kinase receptor family, of which trk is a member, contains other members known to cause cell division when stimulated. Receptors cause cells to divide by activating messengers within their cells. Molecular messengers have the usual inscrutable biological names, such as GAP, Raf, and Sis. The main point is that the kinase receptors evoke cell actions by stimulating other intracellular signals in assembly line fashion. In theory, the trk receptor could stimulate different production facilities at the same time, and this would lead to different effects such as survival, fiber growth, and connectivity. Alternatively, a single assembly line could activate all of these processes.

Enough work had already been performed on kinase receptors outside of the nervous system to identify the many messenger signals they activate. Each signal was a potential mediator of the neuronal effects that NGF elicits by binding to trk. These intraneuronal signals were substrates for the kinase enzyme receptors. The substrate messengers constituted a garden of delights of new molecules with exotic names, including PLC-γ, Raf-1, Ras–GAP, and PI-3 kinase. Each operated its own assembly line within cells but also intersected with the other assembly cascades at critical junctions. There were potentially enough molecular signal pathways flowing from a receptor such as trk to trigger the diverse actions resulting from NGF. Were any of these relevant to NGF, trk, survival, and circuit formation?

Signals, Cascades, and Bucket Brigades

The trk receptor autophosphorylation caused by NGF binding initiates a sequence of events leading to life, death, or neuron growth. Autophosphorylation allows adaptor proteins to bind to the inner segment of trk. The adaptor proteins are the first workers in a bucket brigade assembly line that leads to a neuronal response. These proteins have the now-accustomed impenetrable tags such as SHC, yes, SH-PTP1, and PLC-γ. In general, each adaptor is the initial member of an independent bucket brigade cascade that passes along a signal ultimately leading to a critical effect, such as survival.[4]

We can follow one of these parallel signal cascades, from NGF binding to biological action, to capture the transformation of molecular signaling into neuronal survival. NGF binds to trk and elicits the addition of phosphate molecules to different parts of the intracellular segment of trk. Adding phosphate to one site attracts the adaptor protein, SHC (pronounced "shick"). This begins a sequence of events in which a signal is passed along proteins in the cell. In this cascade, each protein biochemically changes the next protein in line, and this leads to the signal's transmission. Beginning with SHC, the following proteins are modified in sequence:

$$SHC \rightarrow Grb2 \rightarrow Gab1 \rightarrow PI3K \rightarrow AKt \rightarrow p70rsk$$

The cascade results in nerve process formation and survival of the neuron. This is the bare bones skeleton of a cell survival/fiber growth assembly line, as far as can be deciphered. The exact details of each signal transfer and the way it finally ensures neuron survival have yet to be elucidated. But the general game plan is beginning to emerge.

Many signal pathways fan out from a single trk receptor. Different pathways lead to different biological effects, such as larger neurons, altered neuron shapes, fiber growth, and survival itself. In this manner, the binding of NGF to a single receptor molecule can result in manifold biological effects.

This overview is a gross oversimplification. We do not yet know whether there are hidden orthodox relations among signal pathways and their cellular effects. Do elements of one path influence another? We are not yet aware of subtleties of NGF-trk interactions that may favor one or more pathways. Do trk receptors on one part of a neuron activate the same signals as those on another part of the neuron? Do trk receptors on immature neurons activate signals on mature and aged neurons? And what about time? Does the duration of NGF-trk binding differentially activate pathways? Does speed of the bucket brigade affect the final biological responses? Do bucket brigades operate over different time frames, yielding their diverse effects? We seem to have learned so much, and know so little.

Now that we have a rough sketch of how the first known survival factor affects neurons, we might reconsider one mildly inconvenient fact. All the elegant experiments that we have described, all the powerful methods that were employed, all the ingenious interpretations that we have developed, all the detailed insights provided, all the predictions to which we are now heir, are based on studies of peripheral neurons or cancer cells. Not one experiment was performed with brain neurons. What does all of this have to do with the brain? With the mind? Read on.

❖

The index cards that Enoch now uses to negotiate the challenges of work do not protect him from episodic, alien, abusive outbursts, momentary confusion, or waking at night disoriented and panicky. Progressive brain dysfunction produces cognitive and emotional deficits, but precise mechanisms remain a mystery. After decades of futile, frustrating search, hints begin emerging from several laboratories that nerve growth factor may act on brain neurons. Experiments performed in brain cell cultures and in living animals provide the first fragmentary bits of evidence. Unexpectedly, brain neurons that prominently degenerate in Alzheimer's disease are rescued from damage and death by the factor. As always, however, the discoveries raise more questions than answers.

❖

3

HOW TO LIVE IN THE BRAIN

❖

ENOCH WALLACE, in desperation, discovered the value of index cards and familiar surroundings. He now depends on cues and context. With index cards, he can retrieve a colleague's name, delineate the parameters of a new merger, outline the keys to balance sheets, list cash flow, revenue, profit, debt, book value, receivables. Two years ago it was a trivial matter to carry it all in his head. Now he jokes about needing a "peripheral brain." But, even with the cards, the facts simply do not flow as fluently. And certainly new facts, new balance sheets, new company structures are simply not assimilated, as before. Often, in a new situation, Enoch is momentarily confused. But he knows how to stall, ask questions, recover his bearings, and take notes on the cards. If only he could concentrate again. What with his mother's fractured hip, his daughter's impending wedding, and all the dinner parties, he doesn't pay attention at briefings; his mind is elsewhere.

He has had several frightening instances at home, waking in the dark night, disoriented. It was very different from the usual temporary disorientation he had known in the past—the confusion that clears in a moment as he moved from sleep to wakefulness. The recent episodes cleared only after his wife turned on the lights. By then his heart was racing, he was drenched with perspiration, he felt short of breath.

There is no easy divide between the cognitive and emotional debility of Alzheimer's disease. Enoch Wallace, outdoorsman, had always greeted minor annoyances and semi-emergencies with an equanimity welcomed by camping companions. From broken tent stake to twisted ankle, he could stand back, reflect, and come up with a solution, whether action or the right piece of cared-for equipment. His ability was low-key, and seemed to spring from a well of confidence that was never overbearing. The same, or a related talent, allowed him to teach junior associates at the firm without threatening or discernible impatience. This package of personal traits was taken for granted by his peers, regarded as unremarkable as his height or facial features. Certainly nobody credited him with any special sensitivity or particular consideration. This happened to be Enoch Wallace. Perhaps it was the very lack of background notoriety that attracted attention to several recent episodes. Episodes that would have been routine for Ben Schilling, Marvin Goldfarb, or David Selene, hot-shot whizzes at the firm, who suffered neither fools nor critics gladly.

Joan DeLise, a recently minted business school graduate, had submitted a report on a company to Enoch, complete with routine financial analyses, and concluding with projections and recommendations. For this high tech company, however, she neglected to consider competition from a related industry segment in video conferencing. Enoch called Joan into his office and launched into a diatribe. "How could you possibly ignore the impact of video conferencing? And what are the implications of the endless wireless satellite capabilities that will be bursting on the market? Joan this is simply inadequate." Worse than the condemnatory words was the tone. This Enoch Wallace raised his voice to a pitch that was audible to those in the carpeted corridor outside of his office. Joan left the conference confused, angry, and near tears. Ben and Marvin were simply dumbfounded. After several seconds, Enoch was dumbfounded as well. What had happened? Enoch found Joan, offered a half-apology, and the whole incident was forgotten, almost. Enoch was left with a residue of unreality. He found himself more confused than contrite.

The next day, at a small conference in the boardroom, considering the outlines of a negotiation strategy, Enoch became uncharacteristically argumentative. His usually probing, thoughtful questions concerning the projected earnings for a target company, Circor Partners, became a full-blown critique. He actually accused Ben of doing a "sloppy job" collecting the data, and stormed out of the room, leaving a group of bewildered associates in their seats around the table. Later, in the coffee room he mumbled something about his mother in the hospital and his daughter's upcoming wedding. Enoch's partners were only too happy to drop the whole thing, and ask no questions. There was a sense of a conspiracy to protect Enoch from something, although no one was quite sure from what.

For the small society of the firm, a team with ties that extended back for a generation, the sense of shared destiny was a driving, cohesive force and Enoch's friends were apprehensive. It was becoming increasingly difficult to cover up for Enoch. His acts, his performances were more and more flagrant. Enoch, the reserved, articulate, yet understated repository of wisdom, was now beginning to pontificate.

❖

Brain signals govern function of brain and mind. As the agents of communication, signals are necessary for virtually every aspect of brain output, from thought to emotion to motor action. Signal derangement, not surprisingly, results in a wide variety of neuropsychiatric disease, from depression to dementia to Parkinson's disease.

Jeanette McCready was legendary for her hard work. At age fifty-two, she had raised six children, including twins, was a constant carpenter companion to her husband, constructing and endlessly improving their house, and worked at the bar, nights at the club. Her constant physical exertions made her particularly aware of the routine aches, pains, bangs, and bruises of life's labors. She was attuned to her body, without being preoccupied, but recognized well that fifty-two was not thirty-two.

Framing out a new guest room adjoining the living room, she experienced a new type of fatigue after four hours of nailing up several walls of studs. Her right arm felt peculiarly stiff, almost rigid. It was different from the usual muscle aches, but she attributed the new sensation to the exertions of an aging, abused body. Over the ensuing months, the stiffness actually seemed to increase, and she couldn't work it out by massaging her forearm, stretching her fingers, or flexing her muscles. In September, when Jamie and Ian returned to school, she first noted a shaking, a coarse tremor of her right arm. The shaking progressed rapidly, and by Christmas, her hand and forearm shook vigorously enough to interfere with writing and pouring drinks. At that time she also experienced stiffness in her right leg that interfered with walking. She developed a strange, stiff-legged gait: Jeanette took small, stiff steps, had difficulty maintaining her balance, and tended to lurch forward and to the left.

By July 4, other difficulties had arisen. She was definitely moving more slowly altogether. She noticed, in particular, that she had problems arising from a chair, of all things. Accustomed to leaping up to begin a job her whole life, it now literally took several seconds, and an act of will, just to stand up and begin walking.

Rigidity, tremor, slowed movement, abnormal gait. The hallmarks of Parkinson's disease. The disorder is attributable to death of a group of neurons deep in the brain that prevents the delivery of a critical brain signal to other, target neurons, which control aspects of muscle movement. The signal is the neurotransmitter, dopamine. The dopamine brain signal controls motor "programs" that are deranged in Jeanette. Treatment with L-DOPA replaced the dopamine signal in Jeanette's brain, alleviated the symptoms, and allowed her to live her normal, taxing, life of labor.

Brain signals, hundreds, perhaps thousands, mediate communication and underlie motor activity, sensation, perception, and mentation. It was critical to determine whether nerve growth factor, or NGF, was a brain signal mediating life, death, and connectivity.

❖

We endow present circumstance with such insistent inevitability that it's often difficult to realize that it did not have to be this way. Was it preordained that I would sleep on the left side of the bed, be right-handed, have coffee with muffin at 6 A.M., preview each day with my son, Reed, drive the same central Jersey route bathed in the rising sun, past horse farms to the medical school laboratory? Where was it written that your family, home, friends, job, were to be just so? We hover between incomprehensible accident and apparent predictability; science is no different.

The early work on NGF in a tiny bedroom in Rome, with Brownshirts in the street and Il Duce on the podium, need not have led anywhere in particular. The nerve growth factor story has the twists, turns, dead ends, misunderstandings, inappropriate assumptions, and failures, alloyed with occasional insight and happy accident, to represent our species' relentless need to understand. For nearly half a century, there was little indication—despite valiant, often misdirected efforts—that the saga would lead to brain and mind.

Perhaps it was the very drama of the early experimental results that led gifted scientists astray. Adding NGF to cultures of sensory neurons, which convey pain, temperature, and touch, or to sympathetic neurons, key parts of the fight-or-flight system, changed them utterly. Most dramatically, neurons that would otherwise have died, were rescued and survived. The neurons grew in an astounding fashion and sprouted long fibers that connected to targets. The results were reproduced in living animals; hence, this was a fact of life, not a "tissue culture artifact." All of these miraculous NGF actions affected peripheral neurons outside the brain. Sensory and sympathetic neurons are strategically located in the body, roughly adjacent to the spinal cord, and are not part of the central nervous system (CNS), which is confined to the brain and spinal cord.

Surely, though, NGF is the way to build a brain. NGF ensures that neurons survive, and it induces the neurons to grow fibers and form connections that constitute pathways and circuits. Communication along the pathways gives rise to mind. Most perspicacious scientists knew that the next task was to demonstrate how NGF acts on brain neurons.

NGF, the Brain, and Scientific Revolutions

Moving experiments from peripheral neurons to the brain is akin to changing universes. The peripheral nervous system (PNS) consists of tens to hundreds of thousands of neurons. Peripheral neurons receive input from their central counterparts and send fibers to targets such as the heart. The brain, by contrast, is a nearly uncharted galaxy with 100 billion to a trillion neurons—nobody knows for sure. About a million times as many neurons are in the brain as in the entire periphery, each with 1000 to 10,000 connections to other neurons. By conservative estimates at least 10^{15} (1,000,000,000,000,000) individual connections are in the brain. These vertiginous numbers have painfully practical consequences for those who would study the brain. You had better pick the right neurons to study. If chance is relied on to find the brain neurons that respond to NGF, the experimental project is likely to take hundreds of centuries. In the research-funding world, it's unlikely to get a grant of that duration.

How can the appropriate brain neurons be chosen from the constellation of possibilities? The only conceivable clue is the identity of the peripheral neurons that respond to NGF. Researchers reasoned that the brain neurons that most closely resemble the NGF-responsive peripheral populations should be examined. Alas, neurons, like people, exhibit a multitude of characteristics; any one of them can reveal a family resemblance. Neurons can be classified by size, shape, length of fibers, number of fibers, form of fibers, target connections, brain location, electrical properties, biochemistry, function (sensory or motor), embryonic origin, and so on. The number of potentially relevant traits is endless. How, then, did scientists pick among the trillion or so neurons in the brain?

The choice was strongly influenced by the neuroscientific revolution in the late 1950s through the early 1970s. It had long been known that neurons use electrical impulses for signaling. Evidence was beginning to emerge that neurons also communicate by sending chemical signals.[1] Julius Axelrod at the National Institutes of Health in Bethesda, Maryland, Ulf von Euler at the Karolinska Institute in Stockholm, and Arvid Carlsson at the University of Göteborg, also in Sweden, were leading the charge. Their laboratories were discovering that sympathetic nerves use the hormone noradrenaline to communicate with targets. Evidence was mounting that sympathetic nerves make the heart speed up, the blood vessels contract, and the iris dilate by releasing noradrenaline (called "norepinephrine" in the United States). Von Euler, the latest genius in a distinguished Swedish scientific family, localized noradrenaline to nerves and demonstrated its release as a signal. Axelrod, the prototypical American who worked as a laboratory helper (he did not receive a doctorate until age forty), discovered that injected norepinephrine became localized to sympathetic nerves; he defined the biochemical reactions by which nerves synthesize and metabolize the signal. Carlsson was discovering that similar signals might also work in the brain. The electrically wired nervous system was turning into a vast network of chemical signals. The very language of neuroscience was changing from electrodes and millivolts to molecules and enzymes. Laboratory chats, national conferences, and international symposia were dominated by talk of the new neurotransmitters. These insights led to a Nobel Prize several years later for Axelrod and von Euler.

Against this background, another revolution exploded. Attention was drawn to noradrenaline in the brain. A new method directly visualized noradrenaline in nerves. Using formaldehyde vapor, Bengt Falck and Nils-Åke Hillarp at the Karolinska converted invisible noradrenaline into a shimmering, fluorescent green.[2] Under the fluorescence micro-scope, noradrenaline-containing neurons emitted a hypnotic, other-worldly green fluorescence. By gazing through the microscope in a darkened room, they could see neuron cell bodies, elongated fibers, and nerve terminals in garish green detail, against a black background of other cells and neurons that did not contain noradrenaline.

The Falck–Hillarp fluorescence technique gave scientists their first glimpse of neuronal signals. The work of Axelrod and von Euler pre-pared everyone for the sight of peripheral sympathetic nerves aglow with green fluorescence. But in the darkened microscope room, looking at the brain was like gazing at the heavens for the first time. Like whirling galax-ies, green fluorescent fibers appeared to be everywhere. Discrete path-ways could be followed in minute detail; cell bodies of origin were mapped and located in specific brain areas.

Populations of brain neurons used the noradrenaline transmitter just like peripheral sympathetic nerves—a shared neurotransmitter trait that caught the attention of those studying NGF. Since peripheral, noradrenaline-using sympathetics respond to NGF, the brain's fluores-cent neurons were clearly the ones to study. After all, the neurotransmit-ter phenotype of a neuron, the signal that a neuron uses to communicate, lies at the heart of the neuron's identity. This was precisely the simplify-ing strategy needed to investigate the actions of NGF in the brain. Out of the trillion brain neurons, only a few thousand were fluorescent. As im-portant, the fluorescent neurons were localized to two discrete areas of the brain that could be easily identified and manipulated experimentally. The experimental nightmare was not going to materialize: it would not be necessary to search through a trillion brain neurons to identify a po-tentially tiny NGF-responsive population. Though unspoken, though never subjected to critical debate, relief rippled through the scientific community and a fateful, tacit consensus was reached. The fluorescent brain neurons should be the focus of attention.

The fluorescent brain neurons use noradrenaline, adrenaline, or a closely related chemical, dopamine, as transmitters. The group of norepinephrine neurons, lying deep in the brain, is known as the locus coeruleus because it has a blue tinge. The locus coeruleus appears to play roles in arousal, attention, and stress. This tiny group of neurons certainly could exert widespread effects on mood and behavior since it sends fibers to almost every area of the brain and spinal cord (Figure 3-1).

Dopamine neurons were destined to become the focus of intense interest years in the future because their death causes Parkinson's disease.

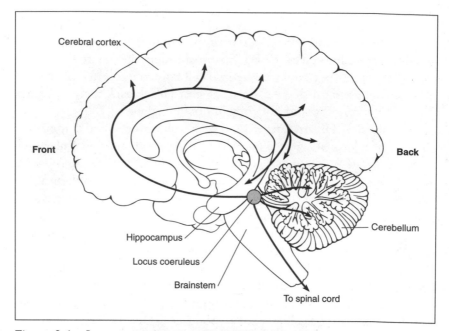

Figure 3-1 Locus coeruleus noradrenergic (norepinephrine-containing) neurons in the brainstem. In this schematic, the cerebral hemispheres have been divided and we are viewing the middle (medial aspect) of the hemisphere. The locus coeruleus, consisting of only several thousand nerve cell bodies, sends fibers throughout the central nervous system. These nerves innervate the cerebral cortex, the cerebellum, and the spinal cord. Consequently, they are positioned to influence such global states as arousal, attention, and stress.

The neurons control certain complex muscle movements. As a group these neurons are part of the substantia nigra (they contain black pigment) and, like the locus coeruleus, lie deep in the brain (Figure 3-2). So, the black and blue norepinephrine and dopamine neurons became prime vehicles to define the role of NGF in building brain and mind. Experiments proceeded at a furious pace to uncover the actions of NGF on the norepinephrine and dopamine brain neurons. Teams in Sweden, the United States, Italy, France, and Great Britain devised ingenious strategies in culture and in live animals to unravel the secrets of NGF in the coeruleus and nigra. Activity reached a frenetic pitch in the 1970s. Though experimental approaches followed those for the periphery, the brain presented special problems.

The body protects the brain in a throne room guarded by fortresses. Locked in the inaccessible boney box of the skull, the brain is impervious to approach without doing violence to the cranium. The facile approaches to peripheral neurons have no equivalents in the brain.

The body also protects its master organ from blood-borne access. The blood-brain barrier prevents most foreign circulating chemicals from entering the brain, thereby protecting it from exposure to potential toxins. While life-giving oxygen, glucose, and many amino acids are allowed to pass, most, but not all, chemicals that would render us goofy and derange our behavior are denied entry. (One can't help but wonder, however, at evolution's wisdom in allowing alcohol, hallucinogens, and certain stimulants admittance to the controls.)

Thus NGF injected under the skin, into a muscle, or even into a vein is not allowed to enter the brain. Scientists were not deterred. They delicately drilled holes into the skulls of anesthetized rats and mice and injected NGF or NGF antiserum (anti-NGF) directly into the substantia nigra or locus coeruleus areas. They also ran tubes into the skull holes to deliver NGF or anti-NGF continuously. The growth factor or its antiserum was given in single doses, every few hours or daily. Some experiments were run for hours, some for days, others for weeks. Every conceivable dosage schedule was used. The growth factor was administered to newborn, developing, adolescent, and adult animals to ensure that any critical periods of responsiveness were not missed. NGF and anti-NGF were employed in every imaginable combination to detect any potential effects. Nigral and coeruleal neurons were examined for survival, size, fiber growth, fluorescence, and biochemistry; behavior was also assessed to determine whether the growth factor was having special effects.

At the same time, scientists developed methods to grow the catecholaminergic (norepinephrine and dopamine) brain neurons in culture to complement the live animal experiments. Brain neurons are far more delicate and picky than their peripheral relatives. Each group of brain neurons has growth requirements necessitating special media, unique surfaces for growth, and the company of select companion cells. Fastidious brain neurons simply die if conditions are not precisely right. In the last quarter of the twentieth century, with men on the moon, relativity theory in place, quantum mechanics elaborated, and atomic power all-too-evident, brain culture was part science, part art, and part black magic. Only a few laboratories across the globe had the near-mystical expertise to grow brain neurons in culture.

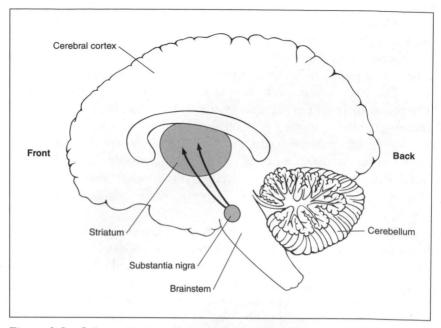

Figure 3-2   Substantia nigra dopaminergic (dopamine-containing) neurons in the brainstem. These neurons send fibers to the striatum, influencing motor programs and coordination. They degenerate in Parkinson's disease.

NGF and anti-NGF were added to the nigral and coeruleal cultures in every conceivable combination, at every imaginable dose, for all sorts of time intervals. Neuron survival, size, fiber outgrowth, norepinephrine or dopamine content, biochemical measures of health, and electrical properties were monitored. The results were completely consistent—and thoroughly depressing. NGF had no effect on locus coeruleus norepinephrine neurons; NGF had no effect on nigral dopaminergic neurons. Anti-NGF had no effect on neurons in animals or in culture.

The conclusion seemed inescapable: NGF plays no role in the brain. Action of the growth factor is a peculiarity of the peripheral nervous system. The brain, the masterwork of biology, must be governed by other principles entirely. It was scientific hubris to have thought that rules governing life and death, and the formation of connections in the simple peripheral nervous system, would apply to the brain's magisterial complexity.

So many different experiments were performed, each with the same depressing negative results, that detailed description would be superfluous. One example illustrates the great lengths, and the subtle approaches that were taken by those so eager to uncover a role for NGF in the brain. Martin Schwab, a scientific star working in mountaineer Hans Thoenen's laboratory (then at the famed Max Planck Institute in Munich) devised innovative studies as the 1970s were drawing to a close. Schwab, a fellow German–Swiss, was a near-perfect complement to the commanding, severe Thoenen. Though no less precise and intellectually demanding than Thoenen, Schwab exhibited the scholar's gentle, considered esthetic. Thoenen's sinewy 6 feet, 5 inches exuded commitment and focus; Schwab's willowy 6 feet, 2 inches was topped by a professor's spectacles under pale blonde hair. Where Thoenen was direct, uncompromising, and even intimidating, Schwab was measured, attentive, and sensitive to any evidence that might change his view. Thoenen was fashioned of the blood and iron required to pursue solitary, high-risk scientific ascents; Schwab was endowed with an openness to new directions, unconventional approaches, and new opportunities.

Rather than attempting a frontal assault by demonstrating a direct NGF action on the catecholaminergic (norepinephrine and dopamine) brain neurons, Schwab used a different strategy.[3] Thoenen and Hendry had discovered years before that NGF in the peripheral nervous system was transported backward (retrograde) from target to nerve cell body (see Chapter 2). Thoenen and Schwab reasoned that retrograde transport should be detectable in the brain, if NGF played a role. They took advantage of the fact that the locus coeruleus sends fibers to its target cerebral cortex, the gray matter that covers the outside of the brain. The cortex, the center of conscious thought, was injected with radioactively tagged NGF; they hoped to detect retrograde transport to the locus coeruleus, which would demonstrate that the same processes which transport NGF in peripheral neurons also operate in the brain. This might be the first step in discovering hidden brain actions.

The results were not simply negative—they were bizarre. NGF injected into the cortex was not transported back to the locus coeruleus at all. The tagged NGF turned up unexpectedly in another group of neurons entirely. Radioactivity collected in neurons at the base of the brain. These neurons use an entirely different transmitter signal, acetylcholine. This group of neurons is collectively called the basal forebrain cholinergic (acetylcholine-using) neurons. The Schwab and Thoenen scientific publication in 1979 encapsulated the confusing futility of all efforts to detect an action, any action, of NGF in the brain. As the Sorry Seventies drew to a close, it was all-too-evident that NGF had no home in the brain. At the very least, a scientific dead end had been rigorously pursued to its conclusion. The search for the secret of brain and mind architecture would have to proceed in other directions.

At the time, the Schwab-Thoenen 1979 publication was often cited as the last straw in a frustrating search. A nearly incalculable number of scientist-years had been devoted to defining a role for NGF in the brain. In the intense world of advanced science, researchers were fed up with NGF. If it is possible for a community to turn against a molecule, NGF was the neuroscientific outcast of the era. In 1980, sane scientists did not utter the words "NGF" and "brain" in the same sentence.

Insight from Alzheimer's Disease

Even so, the basal forebrain cholinergic neurons, in which Schwab and Thoenen had found NGF radioactivity, were becoming interesting for other reasons entirely. The cholinergic neurons were to become a hot topic in the distant clinical world of Alzheimer's disease.

By 1980, there was a growing awareness that society was suffering from an epidemic of alarming proportions: late-life, degenerative neurologic disease. These disorders include Parkinson's, Lou Gehrig's, and the prototype, Alzheimer's disease. The numbers were staggering. In the United States alone, approximately 4 million patients suffered from Alzheimer's, and about 228,000 new cases were diagnosed annually by the end of the decade.[4] Financial wizards estimated that Alzheimer's cost the United States $30 to 40 billion annually for custodial care alone. The cost in terms of human suffering was incalculable. Yet, the cause of the Alzheimer's epidemic remained a frightening mystery.

Fragmentary clues were beginning to implicate the same cholinergic neurons at the base of the brain. Hints were emerging from different quarters. Clinical studies documented what laypeople had long regarded as common knowledge: thought processes often decline in the elderly; for example, memory for recent events and experiences is notably compromised. Aged rats, mice, and monkeys, all fellow mammals, share this sad fate with humans. In addition to bemoaning the tragic ways of the world, scientists were beginning to examine the brains of elderly, demented patients.

A series of postmortem studies on Alzheimer's patients detected abnormalities in the basal forebrain cholinergic neurons.[5] An association between loss of mind and loss of normal brain cholinergic neuron function was beginning to emerge. There was more.

In one startling experiment, neurologists treated healthy, young volunteers with a drug that blocked function of the brain cholinergic neurons.[6] The young adults had the same recent memory defects as in the elderly. A battery of tests revealed other mental deficits characteristic of elderly, demented patients. Experiments on animals were in agreement. Blocking cholinergic neuron function in young monkeys produced a striking recent memory deficit characteristic of elderly people.

Using a classical approach, the next question was whether the memory deficits associated with cholinergic blockade could be reversed by treatments that relieve the blockade. Young humans or monkeys were treated with the cholinergic blocker. Memory deficits reliably appeared. The subjects were then treated with a drug that partially relieves the block and memory deficits were remedied. The case was growing stronger. Abnormalities of the cholinergic neurons at the base of the brain were clearly associated with memory impairment.

The evidence was mounting. Brick by brick, a scientific edifice was laboriously being constructed. As so often happens, however, a dramatic, headline-grabbing event moved the issue from undifferentiated background to galvanizing cause. Such a discovery was about to link cholinergic neurons specifically to the plague of Alzheimer's disease. This leap was made possible by the collaboration of a unique group of physician-scientists at The Johns Hopkins Medical School.

The collaboration was emblematic of the convergence of neurology and psychiatry as the emerging field of clinical neuroscience. Joe Coyle, already a renowned neuroscientist, barely 40, also happened to be a clinical psychiatrist who was leading the developing field of biological psychiatry. As a medical student at Hopkins, Coyle's imagination was recognized by Sol Snyder, discoverer of brain opiates, and Snyder became his mentor. After graduation, Coyle joined Julius Axelod's laboratory at the National Institutes of Health, where he defined biochemical aspects of brain development and became a pioneer in the study of brain neurotransmitters. He later studied acetylcholine in neurons at the base of the brain and gathered evidence that the neurons send fibers to the cortex, the center for learning and memory. Coyle's discoveries were replacing id, ego, and superego with neurotransmitter function, brain pathways, and biochemistry.

Mahlon DeLong was a Harvard-trained neurologist who discovered how certain muscle movements are governed by the brain's electrical activity, work that would open up new approaches to Parkinson's disease years in the future. He also characterized patterns of electrical activity generated by cholinergic neurons at the base of the brain.

The third member of this unique team was Don Price, a qualified clinical neurologist and neuropathologist. Don was not only a gifted clinical neurologist, he was also a master at microscopic examination of the human nervous system to determine the pathologic processes leading to neurologic disease. His main interest was degenerative brain and spinal cord diseases, of which Alzheimer's was scourge and prototype.

This team, with their combined expertise on Alzheimer's and on the mysterious neurons at the base of the brain, tackled the problem head-on. They decided to examine the brains of patients dying of Alzheimer's disease and compare their brains with those of patients dying from other causes. Attention would focus on the base of the brain. The Maryland State Medical Examiner's Office would be their source of the human brains. After all the necessary ethical and legal clearances, and the mountain of paperwork, the study was ready to begin.

Employing stringent diagnostic criteria, five patient brains qualified as Alzheimer's, beyond any doubt.[5] For comparison purposes, the team chose age-matched brains from adults who had had no signs of mental impairment during life. Examination of the brains was ready to begin. To ensure objectivity, the brains were intermixed and coded by number so the scientist examining the brain had no idea whether it was an Alzheimer's or a control. In another measure of caution, two investigators examined each coded brain independently. These measures were employed to avoid any possibility of unintended bias. Under the microscope, neuron number, neuron density, and neuron size were scored using multiple methods to ensure accuracy.

The results were overwhelming: a 79 percent reduction in the number of basal forebrain neurons in the Alzheimer's brains. Using each independent method of measurement, there was at least a fourfold drop in the basal forebrain Alzheimer neurons. Agreement among the scientists was almost perfect. For the first time, loss of a specific brain neuron population was associated with Alzheimer's dementia.

Functions of the Basal Forebrain Cholinergic Neurons

While the neurons that degenerated in human Alzheimer's disease were the very neurons that had accumulated NGF in the Schwab-Thoenen experiments in rats, no special connection was made at the time. Scientists working with NGF, cell cultures, and rats did not routinely communicate with physicians at the Alzheimer's bedside. The world of NGF research continued on its separate track, and the formerly moribund field of Alzheimer's assumed a frenetic new life of glamor. The discovery of obscure cells at the base of the brain became the cause of the moment.

The Alzheimer-basal forebrain headlines attracted scientists like flies. Along with neurologists, psychiatrists, and clinical psychologists; neuroanatomists, neurochemists, neuropharmacologists, animal behaviorists, and electrophysiologists all flocked to the brain area that might hold the keys to dementia. An avalanche of experiments was performed to define the precise functions of the basal forebrain cholinergic neurons.

The experiments were designed to answer embarrassingly simple questions about this now-infamous group of neurons at the base of the brain. What is the function of these nerve cells? Why does degeneration contribute to Alzheimer's? As always, the nearly naive, easily posed questions are the most difficult to answer. The standard scientific strategy is to sneak up on the question by addressing more tractable pieces of the puzzle. Individual pieces can then be assembled to provide answers. Where do these neurons send their fibers? Where are the targets that receive their signals? Will identification of the targets provide insight to function? What happens to mind function when these neurons are damaged? What happens to the mind when the fibers leading from the neurons to their targets are damaged? There was no dearth of second order questions that might lead to the big picture. The answers were not long in coming.

Neuroanatomists discovered new brain circuits connecting the basal forebrain neurons to a key memory center called the hippocampus (it seems to look like a horse ["hippo"] to the ever-imaginative anatomists). The hippocampus is a critical memory processing area. It had long been known that patients with a damaged hippocampus suffer from amnesia. This was one piece of the puzzle; basal forebrain neurons presumably affect memory by interacting with the hippocampus memory center.

Enter animal psychologists. If the anatomists were right, surgically cutting the fiber bundle running from the basal forebrain neurons to the hippocampus should interfere with memory function. The surgery was performed on rats. The animals were taught to negotiate a maze by using spatial cues in the environment for a food reward. The rats had to remember the route through the maze. Normal rats performed this memory task so quickly it was difficult to follow them. But the rats with surgically cut fiber bundles had no clue where to go. They exhibited the same spatial memory deficits observed in Alzheimer's patients.[7]

Parallel chemical analysis revealed that the critical basal forebrain neurons did use acetylcholine as a neurotransmitter signal, as had been thought. The neurons contained acetylcholine and all the biochemical machinery necessary for synthesis and metabolism. Drugs that blocked acetylcholine interfered with memory; they mimicked the effects of surgical destruction.[7]

The emerging pieces of the puzzle were providing a clear picture. The basal forebrain neurons used acetylcholine to signal the hippocampus, regulating memory. Abnormalities of the basal forebrain neurons, their connections, or of the hippocampus itself caused memory deficits and contributed to Alzheimer's disease. Understanding the inner workings of the basal forebrain neurons became a pressing need. Why do they die in Alzheimer's? What do they require to live and be well? Can they be rescued? How? These questions were central to understanding memory function, Alzheimer's, and the development of treatments.

Brain Cell Culture

To grasp the details of a neuron's inner life, the cell must be isolated from influences of its neighbors and of the rest of the body. The approach that had yielded the clearest view of the molecular details of a cell's life had been to grow it in culture. Cells of interest were removed from the living animal, isolated, and grown in a dish where each component could be controlled. In the closely monitored culture dish, cells could be examined microscopically, biochemically, and electrophysiologically in a manner impossible in live animals or humans.

Brain neurons were notoriously fastidious, unfortunately, and nearly impossible to grow in culture. Only a few laboratories in the world had the expertise, experience, magic touch, and good fortune to even attempt the task. Nevertheless, the stakes were so high that several scientific groups began growing the neurons in culture.

Several laboratories in the United States and Europe did succeed in growing the cholinergic neurons outside the body. Each laboratory had its unique approach. Culture techniques were passed down from professor to staff scientist, from senior postgraduate fellow to junior fellow, from graduating student to entering student, with constant modification. Over a period of roughly five years, established culture methods could evolve in this way into new approaches. The techniques in one laboratory differed markedly from those in another, though both were successful. There was—and is—no single correct approach; but the different approaches yield comparable scientific results, thus progress.

One laboratory that developed an expertise in growing brain neurons in culture was located at Cornell University Medical College in New York City. The Laboratory of Developmental Neurology at Cornell was in the building next to Moses Chao's facilities, where the NGF receptors had been identified and cloned. The laboratory had a long-standing interest in neuronal signals, including chemical transmitters and even NGF. Investigators were intrigued by the potential of the brain cholinergic neurons. Cheryl Dreyfus, a senior scientist in the laboratory, was one of the rare acknowledged world authorities on brain cell culture. Her career began at some distance from the brain culture dish. As a master's graduate student at Cornell University in Ithaca, New York, she became part of a project studying sharks off the Florida coast. Her mentor held a position at a marine biological institute in Florida, and she found herself on deck, hauling sharks on board to record their habits, loves, and behavior. Somehow the behavior of great whites, hammerheads, and makos led to graduate school in neurobiology at Cornell Medical College.

Dreyfus received her doctorate for elucidating the chemistry of neurons located in the intestine that control digestion. To characterize these neurons in detail, she wanted to study them in culture. As a postdoctoral fellow, she worked for a year in the laboratory of Stanley Crain at Einstein Medical College to immerse herself in the mysteries of neuronal culture.

Crain had been culturing neurons for nearly a generation. When he began his solitary pursuit, his culture studies were generally greeted with bemused indifference. Why would anyone want to grow nerve cells in a dish to record electrical activity? It was difficult enough to record electrical activity from nerve cells in their normal locations. To go through all the trouble of removing nerves from the body to place them in dishes befitting bacteria seemed like lunacy. To perform experiments, Crain fashioned apparatus that was a cross between a Rube Goldberg nightmare and a Stanley Kubrick hallucination. Yet, year after year Crain published his scientific findings, which described the electrical activity of nerves grown in his Frankensteinian Petri dishes. Crain's exotic appearance added to his mystique. Though bald at the summit, he was graced with the shoulder-length white hair of a medieval wizard. To complete the improbable picture, Crain was notably inarticulate—his scientific talks largely consisted of "ahs," "ums," and an occasional "ohhh." There was no comprehensible explanation of why he happened to do what he did for a living. Perhaps the scientific world of eccentrics, exotics, and occasional misfits was tailor-made for Crain. He did not bother his fellow scientists, and they neither bothered nor paid inordinate attention to him. Only toward the end of his career did the method of his madness become apparent to his more polished peers. As he approached retirement, the world caught up. We began to appreciate that to characterize the details of individual neurons and their molecules, culture was the way to go. As we enter the next century, brain cell culture has become an indispensable approach and has yielded unanticipated insights. It is nearly impossible to imagine a time when it was on the fringe.

Stanley Crain also happened to be one of the gentlest, kindest, most selfless members of the species. He was a wonderful teacher at the laboratory bench because he instructed by doing. If Crain was the master mentor, Dreyfus was the supremely gifted disciple. Indeed, she had something extra: Cheryl Dreyfus could make virtually any brain area or cell type grow in culture. Though she began with cells of the gut, she ultimately would culture any cells of the brain that drew her interest. In the process she developed new culture methods that provided undreamt of views of the inner workings of brain neurons.

Years later at Cornell, Dreyfus developed methods for growing brain cholinergic neurons. Guided by her incomparable green thumb, the neurons adapted to culture, sent out long fibers, and exhibited the biochemical traits characteristic of acetylcholine-using neurons. The neurons were easily (for Dreyfus) identified by staining them with colored antibodies to acetylcholine enzymes; the neurons turned the appropriate color and were completely distinguishable from all other brain cells in the culture dish. The colored neurons were visually dazzling. They held the hope of delving into the mysteries of memory and the tragedies of Alzheimer's.

Help from Amazonia

A rush of discoveries was about to unfold. Looking back now, different members of the Cornell laboratory have different views of the events, but the narrative line has commonalities. As the cholinergic neuron project was being initiated, a new graduate student joined the laboratory: Humberto Martinez from Venezuela. An unusual student who had already obtained a medical degree from the University of Zulia in Venezuela, Martinez took one look at the cultured cholinergic neurons and experienced love at first sight: He had a doctoral thesis project! To his newfound love, Martinez brought special qualities that derived from both nurture and nature.

Upon receiving his medical degree, as was customary for Venezuelan medical graduates, he had spent three years deep in the Amazonian bush caring for natives of the rain forest. His 500 slides attested to the size and ferocity of the crawling, flying, stinging, biting insects of the rain forest. He was a favored target, spending extended periods with welts, wheals, bumps, and bruises. Martinez loved his patients but had less affinity for creatures entomological. One fateful, pitch-black, rain-soaked night, while under siege by a squadron of pterodactyl-sized mosquitoes, he made a decision: he would obtain a doctoral degree. With his medical degree and exotic background, he was a most attractive applicant to the Cornell neurobiology graduate program in New York City. Having braved the forest wildlife, he was undaunted by the more restricted hazards of Gotham. He brought a keen interest, a broad medical background, and a fascination with degenerative neurologic disease.

Martinez's first order of business was simple: to optimize growth conditions for cholinergic Alzheimer's neurons in culture—a straightforward, empirically informed, trial-and-error affair. Martinez perfected the dissection procedure to obtain neurons from the live brain, tried different protocols to prepare the neurons for culture, used different biological materials to coat culture dishes with the ideal surface for neuron growth, and experimented with different culture media. He also varied the culture time from one day to two weeks to identify the ideal period for study. In addition to developing the best culture paradigm, his standard measures enabled Martinez to become more and more of a brain culture adept. After six intense months under Dreyfus's tutelage, he was one of the few world experts on the growth of the cholinergic neurons in culture.

Martinez could now begin to ask why these neurons degenerate in Alzheimer's. He approached that question by defining the hormones, chemicals, and other additions to the cultures that optimized growth. First, he carefully characterized growth of the neurons over two weeks: the neurons attached well to the culture dish biologic surface, grew in size over days, extended typical neuronal fibers, and assumed the same shape that they do in the brain. Biochemical measurements indicated that the neurons developed the molecular machinery to synthesize and to metabolize acetylcholine, their all-important transmitter signal. All tests indicated that the neurons in culture were behaving as they do in the living brain. The culture system was perfected. It mimicked life in the brain without the confound of a trillion other neurons, 10^{15} connections, and the rest of the body.

Martinez was ready to add the magic stuff of life to the cultured neurons to prevent degeneration, death, and Alzheimer's. Having already used traditional hormones and chemicals in optimizing growth, his gaze turned to the unorthodox and exotic. The laboratory had extensive experience discovering actions of nerve growth factor in the PNS (peripheral nervous system). Martinez thought this would be an ideal initial offering to neurons living in the shadow of Alzheimer's.

Having arrived from three long years in the Amazonian bush, Martinez was, unfortunately, not conversant with the recent sad stories of NGF's failures in the brain. After all, the Amazonian rain forests were not endowed with libraries containing current scientific journals. Senior scientists in the laboratory took Martinez in hand and related how NGF failed every test in the brain. Several scientist-centuries proved that NGF had no actions in the brain. The giants of neuroscience in Europe and the States were in agreement that NGF was powerful in the periphery, but impotent in the brain. Martinez's mentors, Dreyfus included, painstakingly described all the experiments with the brain's neurons, that utterly failed to reveal any action of NGF. These discussions were all part of the education of a graduate student. But Martinez was different; nothing compromised his enthusiasm for NGF. Instead of relenting and simply learning from his seniors, Martinez did something eminently rational. He buried himself in the recently expanded, magnificent Cornell medical library.

Returning to the laboratory fortified by his studies, Martinez reinitiated discussions with his mentors. All the experiments that had failed to discover actions of NGF in the brain used norepinephrine or closely related dopamine neurons because noradrenergic (norepinephrine) sympathetic neurons in the periphery responded to NGF. No experiments had used NGF with brain cholinergic neurons. So far as Martinez was concerned, the critical experiments had never been done. Many an avuncular arm was placed around the improbably optimistic Venezuelan student to persuade him to give up his futile pursuit. Experiments beyond counting had definitively indicated that NGF just did not work in the brain. To reopen the lost cause was beyond reason.

Martinez dutifully listened to all the arguments but stubbornly maintained his view that NGF might be the answer for cholinergic neurons. Humberto was so committed to his idea that, despite the gloomy discussions, he was compelled to take action. He added NGF to some cultures and used others as controls. Following 11 days of exposure to NGF he examined the cultures. After all the reading and discouraging conversations, he could barely believe the results. NGF increased cholinergic (acetylcholine) function sixfold, 600 percent, in the neurons.

When he excitedly reported his results to his mentors the next morning, he was greeted with thorough, unmitigated disbelief. The consensus was that this most certainly was a mistake, an experimental aberration, an artifact. Undaunted, Martinez repeated the experiment again, and again, and again. If anything, the results were confirmed, becoming clearer and clearer with each experimental repetition. NGF appeared, against all reason, against the great weight of recent scientific history, to increase function of the brain cholinergic neurons.

Martinez's mentors remained dubious in the extreme. There was even talk of finding another laboratory, or another career, for him. Perhaps Martinez should return to clinical medicine, in surroundings more comfortable than the Amazon jungle. Surely any student who repeatedly obtained dubious experimental results was a candidate for another field.

Mentors often commiserate when faced with a problem student. The Martinez situation was no exception. Calls were made to the handful of colleagues across the world who had interests in growth factors and the nervous system, to explore the possibility of other positions for

Humberto. In addition, the underground scientific network, which transmits the latest information telepathically, was also engaged. A strange scenario unfolded. Conversations, whether face-to-face, telephonic, or electronic, often began with the present state of graduate students, but invariably lapsed into the latest scientific results. Not infrequently, with great hesitancy, in hushed tones, one of the senior colleagues would admit to strange results. It appeared that in many experiments, NGF was having positive effects on brain cholinergic neurons. A few tentative, qualified presentations to that effect were even delivered at scattered scientific conferences.

The alarms sounded. Martinez was commanded by his supervisors to drop all other worldly concerns and move faster on his scientific project that instantly became, "The Actions of NGF in the Brain." The real revolution had begun, and Humberto Martinez, late of Amazonia, was at the advance point, on the front lines.

The first order of business was to reproduce the results that he had already obtained. Martinez removed the neurons from the brain, distributed them among several culture dishes, and divided the dishes into two groups. One group was grown in normal medium, and the other was grown in normal medium with NGF added. After either seven or fourteen days of growth in culture, the neurons were analyzed. To assess the function of the cholinergic neurons, Martinez measured the key enzyme that synthesizes acetylcholine. The enzyme is named CAT (choline acetyltransferase), for short. Since CAT is present only in neurons that produce the acetylcholine transmitter signal, it is a measure of the health and function of only these cholinergic neurons. NGF increased CAT six-to tenfold in experiment after experiment. NGF did not affect other cells in the dish. In other words, NGF appeared to act only on the critical cholinergic neurons.[8]

In one set of critical control experiments, an antiserum directed against NGF blocked the actions of NGF, indicating the specificity of the effects. The high degree of specificity implied that NGF might play a role in the living animal; an unexpected result bore this out.

One set of cultures was treated with anti-NGF alone, without added NGF. This was one of many control groups examined. Surprisingly, anti-NGF alone decreased CAT, the measure of function of the cholinergic neurons. Why did the antiserum reduce CAT if no NGF had been added to the cultures in the first place? There was only one plausible explanation. Some of the cells in the culture dish must have been producing NGF. The anti–NGF neutralized NGF that was *normally* produced by brain cells. NGF *normally* stimulated the cholinergic neurons. This unanticipated result implied that NGF is, after all, produced in the brain.

CAT was measured biochemically in the test tube. The next step was to look at the neurons under the microscope and to visualize the effects of NGF, if possible. The neurons were identified by using antibodies to CAT. The antibodies bind specifically to the enzyme only in the cholinergic neurons. By tagging the antibodies with pigmented molecules, the colored neurons are easily distinguished. Again, the neurons were grown in culture for days, one group without added NGF, the other with NGF. The effect of NGF on the microscopic appearance of the neurons was stunning. NGF massively increased the number of neurons stained with brown antibody. In the control, non-NGF culture dishes, Martinez (and, by now, his wildly enthusiastic mentors) searched diligently to find the few faint, light-brown cholinergic neurons; but in the cultures exposed to NGF the sea of brown cells was so dense that it was difficult to see anything else.

In subsequent experiments in other cultures, Martinez examined other characteristics of brain cholinergic neurons. Instead of measuring CAT, which synthesizes acetylcholine, he measured an enzyme that metabolizes the transmitter. NGF also elevated the metabolic enzyme. In sum, NGF affected multiple aspects of brain cholinergic neurons.

Martinez's simple observations provided a surprising amount of information and raised interesting questions. Since NGF dramatically enhanced both enzymes, its effects on the cholinergic neurons were generalized. NGF did not increase just one type of molecule in the Alzheimer neurons. This straightforward finding implied that NGF stimulated the general health and well-being of these at-risk neurons. It was difficult not to dream of NGF-related treatments.

There were additional implications of Martinez's discoveries. NGF did not simply magnify any old personality trait of the cholinergic neurons. NGF increased the very machinery that produces the acetylcholine signal. And, from all the studies performed on humans, monkeys, and rats, the acetylcholine signal itself was known to be important for memory function. NGF itself might be expected to improve memory in living animals through this action.

The initial scientific articles, reporting that NGF stimulated brain cholinergic neurons, were greeted with skepticism, to put it mildly.

Some elders in any community, as guardians of orthodoxy, dutifully attempt to impede progress and fight revolution. The scientific community, just one of many societies organized by the species, is no exception. The work was ignored by most and criticized by others. Scientists are certainly well schooled in the art of creative criticism. There was no shortage of potential problems that may have yielded misleading results. Many of the experiments from different laboratories were performed in culture, and neurons in culture may simply respond abnormally. This is the traditional "culture artifact" argument: "what is the relation of life in a culture dish to life in the real world of the brain?"—an entirely valid concern. There was more. Most studies measured only one or two characteristics of the cholinergic neurons, such as the acetylcholine enzymes. Perhaps these selected measurements did not adequately reflect the health of the neurons anyway. Perhaps the NGF employed was impure, containing a contaminant that simply made the neurons healthier; this is the "it's not an NGF effect anyway" argument. Impurity is always a valid concern. Perhaps the NGF was pure, but it affected other cells that stimulated the cholinergic neurons indirectly. The potential objections ran on and on. Critical scrutiny is central to science, yet the wellspring from which indifference and criticism flowed were the weight of history proving that NGF had no role in the brain. That historical weight was about to collapse under a tidal wave of scientific evidence.

This tsunami had begun as a trickle in a peculiar form of cell culture. In this "aggregating cell culture," the whole brain is dissociated into its vast, heterogeneous component cells, which are then placed in a dish and allowed to grow together. Every imaginable brain cell is liable to be present; it is nearly impossible to determine which cell is which. Nevertheless, in the early 1980s, NGF increased cholinergic enzymes in this complex system, raising the possibility that somewhere in this chaos of cells, cholinergic neurons were affected. Schwab and Thoenen, for all their disappointment with the retrograde transport experiments, had not given up either. Driven by Franz Hefti, another energetic scientist in the laboratory, they treated newborn rats with NGF and found that the CAT (or, as they called it, ChAT) acetylcholine enzyme increased when whole brain measurements were performed. This implied that NGF affected brain cholinergic neurons. The group had not examined the basal forebrain cholinergic neurons, however, and was suitably cautious, if not thoroughly disbelieving. The title of their published article reflects their doubts: "NGF-mediated increase in choline acetyltransferase (ChAT) in neonatal rat forebrain: evidence for a physiological role of NGF in the brain?"[9] Despite the barely concealed skepticism of the authors themselves, Hefti and Schwab carried the campaign forward. Their experiments in culture of the basal forebrain neurons also indicated that the cholinergic enzyme increased.

As part of the campaign, Schwab returned to the issue of retrograde transport of NGF in the brain. He discovered that NGF was specifically transported from the cortex to the basal forebrain neurons.[10] This would be expected if NGF normally played a role in the basal forebrain neurons. These findings, not yet graced with the imprimatur of "discoveries," were complemented by studies in California. Bill Mobley, at the University of California, San Francisco, another former student of Eric Shooter and a longtime contributor to the NGF field, took a different approach.[11] He and his coworkers injected NGF into the brains of infant rats and found that CAT increased in another area of the brain that contains cholinergic neurons, a muscle movement center known as the striatum; so, NGF might affect many different cholinergic neurons in the brain. All these studies, in conjunction with the Cornell/Martinez experiments,[8] "suggested" (to use a noncommittal scientific word) that NGF had an effect on the brain after all.

Scientific progress within the community of practitioners can be divided into discrete stages, though real life is rarely quite so neat, clean, and antiseptic. The first stage of discovery is greeted with skepticism, deep doubt, and a general sense of nonacceptance. Central ideas are not taken seriously, submitted scientific articles have a difficult time being published, and a research grant application on "brain NGF" in 1983 would not have been funded by any sober scientific panel. During this stage, the community hews to the line of orthodoxy, and a new idea is relegated to the fantasy fringe of unacceptability. But scientists, unlike many others, are adaptable. As evidence piles on evidence, more and more practitioners participate, and the unorthodox becomes the rule. Scientists are often astonished by the rapidity with which the shunned becomes the rule, which overnight becomes old hat. During the last stage of selective amnesia, the community avers that "it could never have been any other way . . . of course NGF plays a role in the brain." In 1985–1986, the whisper that NGF might have something to do with the brain was on the verge of turning into a roar. Laboratories across the globe were taking note. A flurry of studies was about to flood the scientific journals, a sure sign that the stage of discovery was transforming into the next stage of acceptance.

The second stage of experiments fell into broad groups. The first pursued the culture approach and elucidated the cellular and even molecular basis of action of NGF on the brain cholinergic neurons. From experiments using a variety of culture methods, a consensus emerged that NGF worked directly on the cholinergic neurons and did not require intermediaries.

Another group of experiments, performed in the living brain, was designed to elucidate the effects of NGF treatment in live animals. NGF could rescue previously damaged cells from death, which raised the possibility that NGF might be useful in rescuing neurons in Alzheimer's disease. This objective was the Holy Grail of Alzheimer's research. One dramatic group of experiments even indicated that NGF could heighten memory in impaired animals. The floodgates were opening.

The final broad group of experiments also involved the living brain but was not primarily directed to treatment; rather, they were designed to determine what NGF has to do with normal brain function. Is NGF normally present in the brain? If so, where? Does NGF govern basal forebrain cholinergic function? How? Laboratories from Sweden to Germany to the United States jumped on these problems and rapidly formulated the blueprint. The NGF gene was active in the cerebral cortex, the center of thought, and especially in the hippocampus, a key memory center. The cortex and hippocampus are targets of basal forebrain neurons. NGF itself appeared to be produced in cortex and hippocampus, as expected. Additional studies confirmed that NGF was transported from targets to the basal forebrain neurons where it increased cholinergic function.[12]

As the last decade of the century approached, NGF found a secure place in the brain. The nervous system did not seem so unaccountably asymmetric, with growth factors regulating peripheral but not central neurons. We were even fashioning the tools to think sensibly about great plagues of humankind such as Alzheimer's disease.

Yet there was a sense of incompleteness, of a work barely in progress. Consider that NGF selectively regulates the function of brain cholinergic neurons and even rescues them from injury. That is extraordinary. What about the unsettling detail of the other neurons in the brain? Of the trillion of them, does each group have its own growth factor? Are there any other growth factors in the brain? Is NGF an anomaly? Does life in its infinite economy judiciously employ a single growth factor to build an entire mind? And, so far as disease is concerned, what might rescue the neurons that degenerate in Parkinson's disease? What about Lou Gehrig's disease?

❖

*No longer able to deny the distractibility, periodic
confusion at home and work, and loss of control, Enoch
and Sally visited a neurologist. While Enoch's motor
and sensory skills seemed normal on physical exam, he
fared poorly on tests of memory and concentration, and
committed errors drawing a clock and addressing an
envelope. Something was desperately wrong. Unwilling
to render a diagnosis yet, Dr. Rudick referred Enoch for
an MRI to visualize the implicated brain structures.
With the dawning realization that nerve growth factor
acts on brain structures, an intense scientific search was
launched to identify undiscovered brain growth factors. A
decade of grueling, round-the-clock experiments finally
identified a new factor that, wholly unexpectedly, bore a
striking family resemblance to NGF. So, far from the
assumed barren wasteland, the brain may contain a
family of factors that govern growth and function. A
worldwide cloning frenzy was initiated to identify new
family members.*

❖

4

FAMILIES

❖

THE UNFORTUNATE convergence of poor performance at work, episodic distraction at home, and a subjective sense of loss of control finally brought Enoch to seek medical attention. His internist listened attentively to husband and wife, took a history, found nothing of diagnostic worth on a brief neurological examination, and referred Enoch to a neurologist.

Dr. Rudick had a precise, antiseptic bearing. Unlike Dr. Finelli, the warm, slightly rotund, avuncular internist, the neurologist, Dr. Rudick, was of a different world. His wire-rimmed glasses and hatpins sticking out of his lapels, his breast pocket ophthalmoscope, and his hip pocket reflex hammer were unsettling, quite unlike the familiar, antique stethoscope that was Dr. Finelli's concession to medical technology. A visit to Dr. Finelli always involved the easy banter of friends catching up on family events and benchmarks. Somehow the physical exam of head, chest, and abdomen, the routine rectal exam, the standard electrocardiogram, the periodic chest x-ray, and even the ever-unpleasant venipuncture were always unexpectedly comforting. There was little of comfort in the office of Dr. Rudick, neurologist. In fact, the office was more like a barren cell than a hospice of diagnosis and healing. The entire office consisted of a sharp-cornered desk with facing, straight-backed wooden chairs, a Spartan examining table, and a far wall screen arranged for some kind of visual exam. It was a far cry from Dr. Finelli's clutter of instrument-filled glass and maple cabinets, of EKG machines and respiratory testing devices, of glass slides and microscopes. Enoch found himself thinking that Dr. Rudick deserved his cold comfort office.

Dr. Rudick was not a master of small talk. He engaged in none of the ritualistic, introductory, familiar babble that the species has evolved over the eons to allow strangers to bond, to engage friend and not foe, to allay anxiety of the alien. Rather, Dr. Rudick announced to Sally and Enoch at the outset that he would "initially take a thorough neurological history, and then proceed with the neurological exam." Sally was impressed with his systematic, logical approach. Certainly, this young, well-trained, technophile would quickly get to the bottom of Enoch's problems and identify appropriate remedies. Enoch was discomforted. Where Sally found protective logic, Enoch found a distant, unfeeling sterility. Enoch felt more like an experimental subject than a patient. But they were here to get rid of his lapses, so he could certainly tolerate these small inconveniences.

"Mr. Wallace, what is your main problem? How long have you had problems remembering where you placed your belongings? When did you first have difficulty finding your way around the house? When did you first have problems finding your way home from work? What exactly do you do in your job? How do you actually analyze a prospective company for acquisition? How do you analyze a balance sheet? Do you have difficulty adding? . . . subtracting? How about multiplying and dividing? Do you use a computer? . . . . a calculator? . . . pencil and paper? Who helps you understand and interpret all the information? How are you getting along with your partners, with your secretary, with the receptionist? Do you get into arguments at work? Are people talking about you at the firm? Who are your main friends? Are any of your partners hostile; are they after your job; are they interfering with your work? Let's get back to your home. How long have you lived in this apartment? Can you generally find your way around the house? Where do you spend most of your time in the house; which is your favorite room; do you know your way around all the rooms? How are you getting along with your wife? Are you able to understand each other; do you argue; do you argue more than you did five years ago; what do you argue about? Do you agree, Mrs. Wallace? Do you still have an active sex life; what has happened over the last five years; is work just a complete drain? Tell me about your children. . . ." The questions came so quickly that Enoch barely had time to think. Dr. Rudick never paused, never stopped for a breath, didn't react and barely acknowledged any single answer before proceeding to the next question. Enoch felt like the accused on the witness stand; and the examination had not even begun.

The physical exam was a novel and exotic experience. Dr. Rudick peered into Enoch's eyes through an ophthalmoscope, tested his olfaction (smelling) with vials of coffee, clove, and cinnamon odors, carried out a detailed evaluation of his strength, and even mapped his responses to pin pricks with hatpins over every square inch of his skin. Enoch endured the poking and prodding with equanimity, but the rest of the exam was just plain frightening.

Enoch was asked to remember the color "red," the number "32," and the name "George Washington." Enoch could remember only one of the three items after five minutes. He had difficulty even remembering the exact nature of the task at hand. He was then asked, rather commanded, to start at 100 and count backward by 3's. Once he finally grasped the idea, he became so confused that he broke off after no more than a minute. Or maybe it was just five or ten seconds. This test of concentration was clearly too much for Enoch. Dr. Rudick recorded the result impassively. As test piled on test, Enoch became more disoriented . . . and despondent. Enoch was then asked about current events. He could recount the president's recent victory in Congress, his blunder in the endless Mideast negotiations, and the recent murder case in California. After a shaky start, Enoch's confidence was building. He had no trouble naming a "comb," a "pencil," a "ring," or a "hammer." In response to commands, he placed his hand on his head, closed his eyes, and stuck out his tongue. Devoid of reaction or expression, Dr. Rudick noted that Enoch exhibited no evidence of receptive or expressive aphasia. His pure language skills seemed to be preserved.

Enoch experienced a strange hesitancy naming the doctor's fingers. He was then asked to imitate the positions of Dr. Rudick's fingers. Dr. Rudick opposed his hands in a prayerful pose. Enoch mimicked that without difficulty. But he had difficulty forming a daisy chain by touching the tips of thumb and index finger of each hand in an interlocking position. He was unable to imitate the neurologist's opposing hand position of touching the tip of one thumb to the pinky of the other. These failures made Enoch feel particularly awkward and inept. Dr. Rudick simply took note, offering neither advice nor criticism nor solace. Enoch made several errors addressing a letter; Enoch even reversed, then corrected, numbers on his drawing of a clock. Enoch sensed that something was desperately wrong.

Sally reentered the office and settled into the chair, expecting a clarifying explanation from Dr. Rudick. Enoch sat in his chair, looking disheveled and anxious, but without apparent expectations. "Can you tell us what the problem is, Dr. Rudick," Sally asked, breaking the awkward silence. The neurologist looked up from his chart and computer, and indicated that there were "multiple diagnostic possibilities, which would have to be investigated further." He allowed that "we will have a much better idea when we get the results of the blood tests and the MRI." Though not unexpected, mention of an MRI thoroughly unnerved Sally. She felt her hands tremble before she fully comprehended the words. Something about the technology, something about the institutional hospital setting of the test assaulted Sally. A doctor's office exam was one thing, subjecting Enoch to powerful computerized magnets in deep hospital vaults was quite something else. She could feel life slipping from a little absent mindedness to the distant high technology of brain tumors, strokes, and unimaginable, crippling infirmities. From some deep well, the fragmented image of an asylum's back ward of inmates flashed before her horrified eyes.

Enoch and Sally arrived at the hospital at 8 A.M. for the MRI. They walked through the brightly colored, immaculate hallways, encountering friendly "good mornings" from passing workers and nurses. They sat in the Radiology waiting area until Enoch's name was called promptly at 8:30, and a technician escorted him to a small locker room where he changed into a gown, storing his street clothes in a secured locker. "Now, don't take your credit cards," joked the technician, "those magnets are strong." They entered the MRI Suite. In the center of a huge fluorescent room, a patient-sized examining table extended from one open end of a gleaming cylinder just large enough to accommodate a supine person. Another technician, who could not have been older than twenty-five, asked Enoch to lie on the table, and explained in a soft, warm, sympathetic voice that she was strapping him in to help him remain still so that they would get good pictures. She covered him with a cotton blanket, "in case it's too chilly in here," and Enoch was strangely comforted. Then, she asked Enoch whether small spaces bothered him. Enoch replied "no," failing to comprehend the meaning of the question or his answer.

She said that they were ready to begin, and the table slid automatically into the cylinder until Enoch was completely enclosed by the circular walls that were no more than 12 inches from his body and face. Though he didn't panic, he did begin sweating profusely and he could feel his heart racing. The disembodied, amplified warm voice of the technician announced that they were ready to begin. She explained that they were turning on the machine, and that Enoch would hear a banging sound that was normal. The banging began slowly and softly, but rapidly increased in speed and became deafening. In his thin gown, strapped to the table under the cotton blanket, Enoch was no longer sure where he was. But he could feel warm tears sliding down his face.

❖

Since the discovery of neurotransmitters half a century ago, we have known that signals belong to families. The hundred or so transmitters that convey impulses from one neuron to another are not simply a random assortment of heterogeneous, unrelated chemicals. The molecules belong to groups that are related chemically, processed in the body similarly, and often exert related biological actions. These relationships have direct implications for the genesis and treatment of disease.

With the identification of the transmitter signals that control the heart and blood vessels in the 1950s and 1960s, new treatments for high blood pressure became available. The sympathetic nerves that innervate the cardiovascular system release norepinephrine (noradrenaline) that increases heart rate, contraction force, and constricts blood vessels, all of which increase blood pressure. Abnormal activity of the sympathetic nerves, with elevated release of norepinephrine, is one cause of hypertension. These insights about norepinephrine signal function led to new treatments.

One natural therapeutic strategy involved blockade of noradrenergic (norepinephrine) function. Scientists searched for drugs that interfered with norepinephrine's actions. One of the first agents to be discovered was reserpine. This drug prevented the storage of norepinephrine in nerve terminals. In the absence of storage, nerves were depleted of norepinephrine, and actions on the cardiovascular system were prevented. As expected, treatment with reserpine markedly lowered blood pressure and blocked hypertension, saving lives in the process. Medicine was emerging from the trial-and-error era of empiricism, and entering a new phase of rational therapy based on scientific insight.

But triumph came at a price. Reserpine also blocked storage of other transmitters that are closely related to norepinephrine, including dopamine, and epinephrine in the adrenal and in brain nerves. There was more. Reserpine even blocked the storage of distantly related transmitters that used storage mechanisms similar to those of norepinephrine. One such transmitter is serotonin.

Not surprisingly, in addition to combating hypertension, many other effects emerged, based on actions on norepinephrine's transmitter relatives throughout the body. Delineation of some of the side effects illustrates the importance of the principle of family grouping of signals. In a percentage of patients, a Parkinson's-like syndrome occurred, reflecting depletion of dopamine in the brain (see Chapter 3). Patients were relieved of their hypertension, only to suffer from slowed movement, muscle rigidity, and difficulty walking. In many instances the antihypertensive therapy had to be stopped because the Parkinsonian symptoms were so severe. For these patients, reserpine's success was a Pyrrhic victory.

In other patients, mental depression emerged. Suicide attempts were not unknown. Depression was most probably caused by depletion of serotonin, and perhaps dopamine and norepinephrine, in the brain. Reserpine blocked storage of transmitter family relatives indiscriminately.

These examples, drawn from a universe of effects graphically illustrate the fact of family groupings of neural signals. This lesson was hardly lost on those working on NGF. As the importance of the growth factor was increasingly appreciated, the question came to the fore: are there relatives of NGF hiding in the nervous system? If so, what are the implications for brain function and disease?

Meanwhile, though many scientists toiled valiantly—often at great risk to their professional respectability—to make the brain safe for NGF, intellectual skirmishes continued. Each experiment met meticulous criticism. Scientific protagonists engaged in disputations large and small. Arguments varied from the empirical to the theoretical. While there was general agreement that treatment of the brain with NGF evoked biological effects, many thought this was not compelling. After all, exposure to a host of foreign substances could affect brain function. Ingestion of alcohol, for example, profoundly altered brain function. Yet alcohol is hardly a normal brain constituent. NGF could just be another foreign substance that happens to alter the brain. The very fact that NGF acted on only a few brain neurons was thought to indicate an anomaly, a strange exception to the rule. The selectivity of action in the brain stood in direct contrast to the widespread actions of NGF in the PNS. Even if NGF played a normal role in the brain, this single molecule, regulating a tiny group of neurons, could not possibly build the whole brain, build the mind.

While the rush of evidence began to overwhelm criticisms, and while the variety of experiments all pointed to a role for NGF in the brain, the questions and criticisms did point up a significant problem: What is the function of a single growth and survival factor in the brain's vast economy? To appreciate the salience of this question, let's reset the context: By what processes do brain and mind develop?

Genes and Brain Development

For clarity, let's focus on the cellular level, though development proceeds simultaneously on multiple levels, from the molecular to the whole brain to behavior. Progenitor cell division is precisely timed in the embryonic protobrain; ultimately it generates an extraordinary variety of neurons and supporting cells. The neurons migrate to their appropriate locations, often following complex routes through foreign environments. Neurons have to survive these obstacles before developing. They have to assume characteristic shapes and take on electrical properties, biochemistries, and neurotransmitter signals to begin communicating. The neurons connect with the brain's trillion other neurons; this includes nerve process outgrowth, precise pathfinding that sometimes covers several yards, identification of proper target neurons, and formation of communicative, synaptic connections. Synapse formation itself consists of multiple subroutines. Derangement of any of them can mean miscarriage, birth defects, mental retardation, or psychiatric disease. This brief description omits intricate mechanisms, details, and synchronies that are at the core of brain development. Yet the unprecedented complexity of brain development is apparent. Where does NGF fit in?

One approach to this question requires that we adopt a neuron's view. How does a single neuron successfully negotiate the labyrinth of development? One solution, long invoked by scientists, was that neurons are preprogrammed to develop normally; that is, each infant neuron contains all the information necessary to live, develop, and be well. From using simple arithmetic we know that this is unlikely.

The entire plan, the blueprint for a cell's life, is in its genes. Each of our cells has approximately 130,000 genes. Of those, about 50,000 are devoted to the brain. But the brain contains about one trillion (10^{12}) neurons, each of which forms 1000 to 10,000 connections. There is simply not enough information in each neuron's genes to direct development. Even if all 130,000 genes were used, not enough specificity and selectivity are in the blueprints to direct the development of one trillion neurons—the genes lack 100 million times the necessary information. If the genes were asked to direct connections (conservatively, one trillion times one thousand), there would be about 100 billion times too little information. The inescapable conclusion is that the neuron has insufficient information to direct its own individual development. Genes alone do not a brain build, so where do the other instructions come from?

Developmental Signals from the Outside World?

One answer is that the instructions come from the world outside the neuron as growth and survival factor signals. NGF could fit the bill, but it is just one molecule. Even its most ardent adherents were not prepared to endow it with billions of bits of instructions. Scientific public opinion self-organized into the usual spectrum, with the standard polar opposites. At one extreme, NGF was viewed as the tip of a growth factor iceberg, with a cornucopia of new factors under the surface waiting to be discovered. At the other extreme, NGF was regarded as a brain oddity of little general significance. The developmental information would ultimately derive from spheres yet to be discovered.

Since argument thrives best in the fertile soil of ignorance, the late 1980s were particularly fruitful. Strongly held views were cogently presented from standard armchairs, unencumbered by inconvenient fact. Many a productive scientific meeting adjourned to the nearest pub for a lively bout featuring the many-factor world versus lonely NGF. One of the most persuasive proponents of NGF as an atypical, one-of-a-kind brain growth factor was another true pioneer of the growth factor world, Ralph Bradshaw.

Ralph Bradshaw never entered a room, and most certainly never a pub, quietly. His booming, Boston, tympanic tenor ensured that conversation was not to be dull that evening. Barrel-chested and burly, with full beard, broad smile, and wit punctuated by recurrent laughter, Bradshaw appeared to have just disembarked from his Nantucket whaler. He was fond of dismissing lightly regarded theories as "amusing ideas" and "unusual concepts." No evening was ever complete without a good-hearted verbal Bradshaw brawl. He was also a premiere protein chemist and biochemist. With the focus of Ahab, the instincts of Queequeg, and the insights of Ishmael, he attacked the molecular structure of NGF. Using painstaking methods, he had completely defined the string of 116 amino acids that made up the NGF protein. His elucidation of NGF's structure helped move the entire growth factor field into the modern era. With knowledge of the structure, the mechanisms of action, the nature of receptors, the relations to similar factors, and biology could be approached. Bradshaw was ranked with Viktor Hamburger, Rita Levi-Montalcini, Eric Shooter, and Hans Thoenen as one of the captains of the growth factor enterprise.

Bradshaw argued persuasively that while NGF might well play an important role in the brain, searching for other brain factors would likely be futile. History is not a bad guide to the future. Rather straightforward approaches had revealed the presence of hormones and growth factors, and families within families of the molecules throughout the body. Yet exhaustive experimental projects combining tissue culture, cell culture, and live animals had failed to provide the slightest hint of other growth factors, or hormones, in the brain. From this negative evidence, Bradshaw customarily launched into a litany of technical arguments to bolster his claim that there were simply no other factors in the brain. At this hour of the evening, alcohol was most definitely swishing around brains, and the discourse widened to encompass politics, world affairs, and the future of the universe. Though issues were never fully resolved, these well-lubricated, high-decibel talks outlined criteria that would have to be fulfilled to define new factors, functions that might be served, and even potential methods of discovery. All against Bradshaw's booming background.

It was characteristic, and perhaps entirely predictable, that after all his well-reasoned skepticism, Bradshaw's laboratory, paradoxically, would make a seminal discovery. The setting for the announcement was the first International NGF Meeting in honor of Rita Levi-Montalcini and Stanley Cohen in 1986. Cohen shared the Nobel Prize with Levi-Montalcini for his work characterizing, isolating, and purifying nerve growth factor; these achievements promptly moved growth factors from the imaginary to the physical, thereby validating a new area of biology. The conference was organized by Bradshaw and Shooter in Monterey, California. Monterey Bay with famed marine aquarium, seemed the perfect setting for the New England salt, Bradshaw. He arranged tours of the aquarium in between scientific sessions, orchestrated now-familiar informal debates in the evening, and appeared to be everywhere at once—from Cannery Row to the wharfs to the conference center.

The great excitement at the meeting was the mounting evidence that NGF does regulate brain function. In a frenzy of lectures, scientists from across the globe presented new findings indicating that NGF affected neurons in culture and in the brain, and appeared to be localized normally in the brain. The NGF receptor also appeared to be in the CNS: the requisite machinery for NGF action in the brain was in place. The picture was emerging in an exciting orgy of discoveries. The ambiance in Monterey was electric. We were on the threshold of understanding brain development, function, and perhaps the genesis of mind, in entirely new terms.

Enter a New Function for an Old Factor

Against the background of brain NGF's triumph, Bradshaw delivered a curiously offbeat talk. Though a leader in the NGF campaigns, he did not talk about NGF at all. He described an experimental system for growing neurons from the cerebral cortex in cell culture. Unlike everyone else, however, Bradshaw and his colleagues did not study the effects of NGF. They added an entirely unrelated factor to the brain neurons in culture. They used fibroblast growth factor (FGF).[1] Not surprisingly, FGF was so named because it stimulated the growth of fibroblast cells, which are connective tissue elements present in the skin and in organs throughout the body. It also causes the growth of blood vessels. Bradshaw and his collaborators had taken their lead from reports that FGF might act on neural cells. Though Bradshaw's talk was stunning, it was virtually lost in the sea of NGF hysteria.

Bradshaw reported that FGF stimulated the growth and survival of cortical neurons in culture.[1] These neurons are derived from the cerebral cortex, the thinking part of the brain. Since the cortex is the seat of conscious thought, any discoveries were potentially relevant to mental function. Bradshaw's experiments approached the modest question of how molecules build the mind. In the absence of added FGF, cortical neurons died with time in culture. After two weeks almost none survived. In dramatic contrast, the presence of FGF helped virtually all of the neurons to live for almost two months. The cortical neurons that generate mind needed FGF to survive, at least in culture.

FGF exerted another effect on the cortical neurons. It caused striking outgrowth of nerve fibers. Bradshaw and his colleagues reported that "The enhanced elaboration of processes resulted in neurons making contact with one another."[1] Here was a new survival factor that also created cortical circuits, which lie at the heart of communication, and potentially thought. Fibroblast growth factor, at least in principle, might be pivotal in development and maintenance of the brain cortical connections. Cortical connections are the substance of communication that generates conscious thought and mind. Though it was unclear which of the many cortical neuronal types FGF affected—many control experiments had yet to be performed—the discovery should have rearranged a few tectonic plates, sending shock waves along the California fault lines.

This was the first demonstration that another growth factor, known to be unrelated to NGF, exerted actions on brain neurons. Not only that, but FGF had originally been isolated from the pituitary gland, the master endocrine organ that lies at the base of the brain. So FGF was a normal constituent of the brain. The discovery was made by the scientist most skeptical about finding any other brain growth factors. In principle, at least, brain neurons can respond to factors other than NGF. It was not lunacy, therefore, to entertain the possibility that FGF, and many other factors, might govern brain development and function. Largely unrecognized, a new era had begun.

New Functions for New Factors

If Bradshaw was discovering new functions for old factors, the ubiquitous Hans Thoenen was searching for new functions for new factors. Though he also employed cell culture to detect actions on neurons, his approach was different. Instead of examining brain neurons directly, Thoenen began with peripheral sensory neurons that classically respond to NGF. He used sensory neuron cultures as a detection system. By adding biologic stuffs and monitoring growth effects, perhaps he could find a "new" NGF.

Thoenen's Swiss countryman, Denis Monard, had made a fascinating, though offbeat, discovery in 1973, one that had never been pursued. Using culture, Monard had found that one tumor cell type made another grow.[2] Cells in culture secrete chemicals into the medium just as cells in the body release chemicals. Monard discovered that the culture medium in which nonneuronal ("glial") support cell tumors were growing prompted other cells to grow. The medium conditioned by the support glial cells caused cultured cancerous neurons, neuroblastomas, to sprout nerve processes. In scientific shorthand, a glial cell's "conditioned medium" stimulated fiber outgrowth from neuroblastoma cells.

Thoenen and Monard joined forces to follow this slim lead. They used Thoenen's test system. Sensory neurons from a chick were grown in culture. To survive, these neurons had to have NGF in the medium. Thoenen and Monard hoped to determine whether the medium conditioned by the glial tumor (glioma) cells contained anything that could support sensory neuron survival. As expected, in the presence of NGF their cultured sensory neurons survived, grew vigorously, and extended long processes. In the absence of NGF, less than 5 percent of the neurons survived, also as expected. They were now ready for the big experiment. Thoenen and Monard grew the glioma cells in culture and let them stabilize over several days. Then they removed medium "conditioned" by the glioma cells over twenty-four hours. They added the glioma-conditioned medium to the cultured sensory neurons, and, in an apparent breakthrough, the glioma-conditioned medium rescued most of the sensory neurons in the absence of added NGF.[3]

It was critical to confirm the results with control experiments. Had they simply rediscovered NGF? Did conditioned medium contain NGF, which rescued the sensory neurons? They added antibodies to NGF with the conditioned medium; the antibodies did not prevent the survival of sensory neurons in response to the conditioned medium. As a control, the antibodies did prevent added NGF itself from rescuing the sensory neurons. So, the conditioned medium contained something other than NGF that rescued sensory neurons. They were on the trail of a new factor. Was it identical to the one found by Monard in 1973? By partially purifying the factor, they saw that it was different, which meant that they might have been studying a truly new factor.

What kind of chemical was it? Was it protein, sugar (carbohydrate), or fat (the animal, vegetable, or mineral of the biochemical world)? In approaching this question, Monard and Thoenen treated the conditioned medium with enzymes that destroy proteins. Destroying proteins in the conditioned medium killed the sensory neurons. The new factor was not NGF. It was not the original factor identified by Monard; it was a protein.

Thoenen continued the project in his laboratory, but he added a decisive new twist. He was joined at the Max Planck Institute in Munich by a young scientist from Geneva, Yves-Alain Barde. Together, they were about to embark on a revolutionary scientific journey. They began by characterizing the details of the glial-conditioned medium (which is abbreviated to GCM for convenience). In particular, they continued to compare the effects of GCM to those of NGF; sensory neuron cultures were used as the test system, as before. They hoped to get a clearer idea of the nature of the active ingredient in GCM, and performed experiments to define differences between GCM and NGF. While NGF supported the survival of neurons only from young chick embryos, the survival effects of GCM continued to rise with embryonic age. In the young embryos, neither GCM nor NGF rescued all the cells. The survival effect of combining GCM and NGF was greater than the sum of their individual effects. These results strengthened the notion that GCM has a survival factor different from NGF.

Barde and Thoenen, and a visiting scientist, David Edgar, took off in a new direction. Instead of using GCM, they added extracts of rat brain to the cultured sensory neurons. The brain extracts mimicked the survival effects of GCM: and brain extract–induced survival was different from NGF–induced survival. In one leap, survival of neurons in the exotic culture dish was transformed from a glial tumor oddity to a property of normal brain. The broad implications were clear: the normal brain has an undiscovered survival factor (or factors).

The Hunt for a New Factor

It is one thing to find that a near-primeval brain's soup contains something mysterious that supports neuronal survival. But it is quite another to isolate and identify the active component. Until the molecule is isolated and identified, the official imprimatur of scientific reality is not fully conferred. Soups simply vary too much from recipe to recipe, and laboratory to laboratory, to enable reproducibility—the bedrock of science. Yet a molecule is the same from laboratory to laboratory, or brain to brain. The charge was to purify the survival factor from the brain, a formidable task.

Purification is a daunting needle-in-a-haystack, trial-and-error affair that can stretch on for years. Injudicious decisions can add months, years, or plain old failure to the effort. A scientist naturally hopes to complete the project in his or her lifetime; at the least, the children, and not the grandchildren, might finish the work. Rough calculations convinced Barde, Edgar, and Thoenen that purification from rat brains would be forbidding. The imagined factor was likely to be present at such vanishingly low levels that thousands of rat brains might be inadequate. They turned to the much larger pig brain, which could be gotten from the local slaughterhouse. Before doing so, they confirmed that the pig brain had the same survival activity as the rat brain.

Purification began with 3 kilograms, about six and a half pounds, of whole pig brain.[4] The procedure consisted of six sequential steps, beginning with the creation of a soup from the brain. Since Thoenen and Monard knew that survival was associated with a protein, they separated proteins from the other soup ingredients and passed them through molecular filters, all the while closely monitoring the sensory neuron survival of the preparation. Different filters separated different molecules according to size and shape, or electrical charge. By arraying the steps sequentially, they progressively purified the soup. Then the triumphant trio isolated a single molecular species from the brain that was responsible for the survival of the sensory neurons.

There is something ludicrous about summarizing this monumental effort in one paragraph. Monard described the GCM survival effect in 1973, and the purification took place in 1982—a decade of round-the-clock effort. To obtain the pure factor from pig brain, the molecule had to be purified 1.4 millionfold; that is, in a haystack of about one and a half million extremely fragile straws, a single straw had to be found and separated. If any of the separation procedures damaged the straw, the exercise failed. And a single straw would not do. Millions of copies of that straw were needed to characterize the straw of interest. Moreover, all the steps in the purification had to be so well controlled and characterized that the ordeal could be repeated at will in any laboratory anywhere. The enterprise ranks somewhere among building pyramids, constructing the Sphinx, and parting the Red Sea.

With the new brain survival molecule purified, characterization could begin. The news was dramatic. Although the molecule was different from NGF, it bore striking similarities.[5] It was about the same size as

NGF and had similar electrical charges. Though perhaps a coincidence, it did raise the possibility of a family relation.

With the new molecule in hand, pressing questions vied for attention. Are the survival actions selective for sensory neurons, or do other populations also respond? Are the actions similar to those of NGF and FGF, or are there differences? What cells in the brain, if any, respond? What cells in the brain produce the molecule? Most important, as any new discoverer, explorer, or parent will verify, what shall we name the new arrival? Eschewing romanticism and baroque excrescences, researchers gave the new molecule a minimalist name: brain-derived neurotrophic factor (BDNF, for short).

Defining responsive neuronal populations, assessing differences from known factors, and understanding functions in the brain are all important, but one issue was paramount: What is the structure of the molecule? Though BDNF was clearly a single molecule, its precise structure and physical identity had yet to be defined. A new molecule's structure helps to define its niche in the context of biology. Structure determines function; structure reveals the relation to other molecules; structure indicates what the factor can and cannot do; structure, at the molecular level, defines body and soul; structure determines how a molecule behaves in environments.

Determining the structure of BDNF was, in principle, straightforward. Since it was presumably a protein, or strings of amino acid building blocks linked together in a chain, the accepted approach to determining its structure entails removing each amino acid from one end of the molecule, like beads on a string. As each amino acid bead is removed, it is identified. Then, the next amino acid in line is removed and identified. Though tedious, Humpty Dumpty can be put back together again in proper sequence after all the beads have been identified. An alternative method chops the molecule into pieces. Each chopped piece is analyzed amino acid by amino acid. By careful bookkeeping, the pieces are reassembled in proper order.

Molecular Family Relations

As the sequence of BDNF amino acids was lined up, the identity of the molecule emerged with stunning clarity.[5] Even before any quantitative analysis, it was apparent that BDNF bore a striking resemblance to NGF.

The two factors, BDNF and NGF, shared a slew of identical amino acids in the same positions along the string. Out of about 120 amino acids, 51 were identical in the two molecules. Obviously BDNF and NGF were part of the same, tightly knit molecular family.

If BDNF and NGF were children in the same family, how many other sibs were there? Where in body and brain did they reside? NGF was not an orphan, after all. The nervous system might not have to rely on NGF alone to build its circuitry. Perhaps the whole family, working together, could build brain and mind.

The two-dimensional string of amino acid sequence yielded other clues. Protein molecules like BDNF and NGF fold up in three-dimensional space to act on cells. The three-dimensional folding is directed partly by amino acids in the chain that link with each other and create kinks, twists, knots, and helices. It is in three-dimensional space, the real world, that the molecule interacts with receptor locks on cells. The same link-forming amino acids were in the same positions in BDNF and NGF; so the factors formed similar three-dimensional structures. This implied that receptors for the two molecules might be related; a receptor family might correspond to the NGF–BDNF factor family. While analysis of the molecular structure continued, BDNF could now be localized in the body and permit insights into actions.

Tissues throughout the body were examined to get a crude idea of the normal roles of BDNF. Localization of a hormone or chemical gives clues to function in that area. Widespread distribution of the factor would imply a generalized function. Mouse tissues were examined initially for convenience; the pig was a bountiful animal for purification, but unwieldy for routine experiments in small laboratories. BDNF was undetectable in liver, lung, kidney, heart, digestive tract, or muscle. These negative findings implied that the factor was not an all-purpose survival agent, acting on cells throughout the body.

The factor was localized in abundance to brain and spinal cord, the elements of the central nervous system.[5] These discoveries alone constituted adequate reward for the decade of arduous purification. The presence of BDNF in the brain and spinal cord suggested that common mechanisms might regulate neuronal survival and circuit formation in these different divisions of the CNS. The brain, which governs everything from vision to thought to consciousness, and the spinal cord, which processes sensation and movement, might use the same molecules and mechanisms to build their systems.

Subsequently BDNF was found to support survival of the cholinergic neurons that degenerate in Alzheimer's, the dopamine neurons that degenerate in Parkinson's, and the spinal motor neurons that die in Lou Gehrig's disease. BDNF was most abundant in a key memory center, the hippocampus. Discoveries several years in the future would implicate the factor in learning and memory. The factor purified from brains of the less-than-exalted pig was to figure prominently in that centerpiece of the intellect, learning, and memory.

Other Family Members?

The localization of BDNF, and the emerging appreciation of its functions all derived from knowledge of structure. But to informed scientific discoverer—observers of the BDNF saga, the structure held all-too-obvious implications. The family relation between the BDNF and NGF structures nearly screamed that other family members were out there, waiting to be discovered. A mad race ensued to find the factors. Laboratories in academia and industry throughout the United States, Europe, and Japan engaged in a frenzy of factor purification and cloning.[6,7]

The cooperative—competing cast of characters could hardly have been more variegated. Louis Reichardt at University of California at San Francisco was yet another molecular biologist-turned neuroscientist.[7] He had made the seminal discovery that the amount of NGF produced by a target controls how many nerves grow to the target. This key finding revealed how pathways, and thus lines of communication, form in the nervous system; the work suggested how the blueprint for information processing in the brain becomes organized. His team was also one of the first to discover NGF in the brain. Just incidentally, Reichardt happened to be a world-class mountain climber with numerous maniacal firsts as a mountaineer. (There may be something about studying the brain that drives people to the edge.) Reichardt did not simply climb the legendary, near-suicidal K2 peak in the Himalayas. He was the first human to climb K2 without oxygen (no other species ever had that idea)! Though medical texts instruct that lack of oxygen is not conducive to brain function, the K2 experience did not adversely affect this eccentric genius. (The movie and original play *K2* was about Reichardt and his high-altitude exploits; for reasons known only to the theatrical world, the real-life character of a neuroscientist became a physicist—the obscure neuroscience of brains is apparently less of a draw than the romantic physics of the atom and cosmos.) His stamina, strength, grace, and judgment above 20,000 feet were legendary. Reichardt would disappear from the laboratory without warning for months at a time, turning up 8000 miles away and 25,000 feet above sea level. His collaborators learned that he was somewhere in Tibet on an international expedition. Though everyone had faith in his mountaineering acumen and good fortune, prudent collaborators insisted that he leave his lab notebooks in up-to-date form at all times—just in case. (Reichardt's unusual style extended to his interviews of prospective trainees for the laboratory. He conducted one interview, for example, in a local graveyard, far from the traditional desk-in-conference-room setting.) Reichardt took one look at the BDNF structure and intuitively knew that other players were lurking in the wings. He immediately set out to clone them.

The hyperkinetic Hakan Persson at the Karolinska Institute in Stockholm was also on the trail immediately. Persson had trained as a molecular biologist studying the immune system, in Sweden and then at Harvard Medical School, before starting the Laboratory of Molecular Neurobiology at the Karolinska Institute. Persson, a novel life form of concentrated enthusiasm, had solved the minor problem of being in more than one place at one time: the places were always in different areas of the laboratory doing experiments. He seemed to be everywhere in the lab at once, treating animals, helping with cell cultures, cloning genes, determining gene sequences, and writing scientific articles, all while urging others on to greater effort. To avoid wasting precious time, he moved a cot into the laboratory, minimizing that unproductive interface between sleep and waking that plagues most of us. Always in demand at international meetings, Persson ensured that time wasn't wasted. He always ferreted out the flights that minimized time away from the lab, thereby maximizing sleep deprivation. With a cot in the lab, multiple experiments going at any one time, telephone to one ear and numerous international collaborations in progress, Persson was the Swedish whirling Dervish of neuroscience. As soon as hints of the BDNF structure became available, his team launched a holy war to find and clone the other factors in the family.

Persson's enthusiasm and brilliance radiated from an unusual physical package. He was a severe stutterer. He walked with a congenital limp. He regarded his infirmities with thorough disregard, drove himself beyond the conception of his physically "normal" friends and colleagues, and exuded a seemingly limitless optimistic cheerfulness. Whether related to his merciless drive and physical disregard or to some other underlying infirmity, Hakan Persson died of a heart attack at 40, in one of the tragedies of neuroscience. He left a legacy of discoveries, devoted, outstanding students, and devastated colleagues and friends across the world.

The fledgling neuroscience biotechnology industry also jumped into the fray. One of the most active young companies was Regeneron, headed by the unique Len Schleifer. Schleifer was part clinical neurologist, part scientist, and all entrepreneur. He was a chess prodigy who as a child joined the Greenwich Village chess clubs on weekends and annihilated the hard-bitten eastern European émigrés. He routinely left in his wake a sad band of chain-smoking, would-be grand masters. He ultimately gave up chess for science and medicine and obtained his medical and doctoral degrees under Al Gilman at the University of Virginia. Schleifer was deep in the discoveries of how receptors on the outside of cells relay signals to the interior, altering cell function and behavior. Gilman, who shared the Nobel Prize for this and related work several years later, noted that Schleifer's imagination was legendary at Virginia. He seemed to hold the world's open record for thinking of more experiments in a morning than most of his fellow students could imagine in an entire graduate career. Schleifer also was blessed with the ability to seduce his coworkers into doing the work he dreamed up. After completing training in clinical neurology at Cornell Medical School, he joined the Laboratory of Developmental Neurology, working with Cheryl Dreyfus, and collaborating with Moses Chao. He rapidly appreciated the clinical importance of neuron survival factors and, additionally, understood their commercial potential as new treatments. He started Regeneron to translate his dreams into reality. One of his first inspired acts as CEO of the startup company was to hire George Yancopoulous, a brilliant young molecular biologist from Columbia University. Yancopoulous was also following the BDNF purification and cloning, and he, too, immediately realized that the NGF family must have multiple members.

The race was joined by workers at Genentech, Thoenen's team in Munich, and more than half a dozen other laboratories around the world. Within three years of nonstop experiments, NGF, the lone factor in the wilderness for decades, was lonely no more. When the smoke cleared, the mammalian neurotrophin family consisted of four members: Neurotrophin-3 and neurotrophin-4/5, which complemented NGF and BDNF in the family portrait.[6,7] Later, Thoenen's team isolated neurotrophin-6 from fish. All were in the brain and PNS, and each exhibited unique cellular localizations, consistent with unique functions. All were closely related structurally.

Some Lessons

Though the roles of neurotrophins have yet to be fully elucidated, the adventure that began in another era in a bedroom in Italy has yielded several lessons. Indistinct outlines of life's strategy for building brain and mind have begun to emerge; hence it's time to pause and reconnoiter because nature may be revealing secrets about brain, mind, and the sources of humanity. Our imperfect perceptions of nature's messages can be articulated: Where there is one growth factor, a family will follow. NGF never was an isolated peculiarity, a freak. Rather, NGF was a prototype of the molecules and mechanisms that create the brain, the center of thought, emotion, and consciousness. In retrospect, what was freakish was our thinking that NGF was a chimeric exception to life's rules. Once NGF met the criteria of reality, relatives should have been sought immediately.

Where one family of factors exists, neighboring families are sure to reside. The NGF neurotrophin family heralded the coexistence of the fibroblast growth factor (FGF) family, the insulin growth factor family (IGFs) and the so-called tumor growth factor (TGF) family, among many others yet to be discovered. Families upon families of factors build and maintain the brain. It is nearly impossible to estimate how many factor families we'll discover. We can be confident that unimagined factors serve functions not yet even conceptualized.

When life finds a strategy, it is used and reused, co-opted and refashioned to build organs, systems, and the brain. If this notion is correct, it provides other hints. Survival, nerve fiber growth, and even the formation of connections may not be the final stories on neurotrophin functions. If life likes a molecule, it may well be used again and again for many functions. With each new discovery of growth factor function, we should be aware that many other unsuspected actions might be hidden from view. As we'll see, trophins operate in mind function in ways that were unsuspected during the frenzy of carnivorous cloning.

The lesson that signals in the cell's environment guide and instruct cell behavior extends to many functions, perhaps to virtually all phases of a cell's life. Since many environmental molecular signals are responsive to experience, distinctions between nature and nurture, that age-old argument, break down. A cell, a neuron, is not enslaved by its genes. The interaction of environment and genes, nature and nurture, influences the personality of a neuron or a collection of them.

Many environmental signals support neuronal survival. Why do neurons need help in surviving? Does the brain hide a threat to life, liberty, and the pursuit of happiness for resident neurons? Is life in the brain so dangerous that it requires the constant vigilance of survival factors? What is the nature of life and death in the brain? Read on.

❖

*Forgetting names of friends, misplacing pens and papers,
exhibiting confusion at the news store and experiencing
spells of staring into space, Enoch now requires an
informal support team, and can only make token visits to
work. Sally's anxiety, insomnia, loss of appetite, and
crying spells are overwhelming the need for constant
protective vigilance around Enoch. There now can be
little doubt that neurons deep in Enoch's brain are dying.
The nature of cell death had occupied scientists for
decades since the original discoveries that death,
paradoxically, plays a central role during development.
Elucidation of the genetics of cell death, and the discovery
that trophic factor deprivation induces cells to commit
suicide has begun to delineate underlying causes of cell
death in brain disease.*

❖

5

DEATH AND THE BRAIN

❖

THOUGH ENOCH had faithfully taken nostrums prescribed by the neurologist for the past six months, he noticed no improvement. His wife grew alarmed by his lapses. Of course, he no longer worked his accustomed five-and-a-half-day week. Through mutual agreement, he visited the office two or three days a week. Participation in merger and acquisition negotiations was out of the question, but he still engaged in behind-the-scenes discussions, and occasionally contributed an acute insight that surprised and delighted his concerned partners. With the near-constant support and intervention of Ruth, his long-time administrator, and Tania, his personal secretary, he appropriately dispensed his reduced responsibilities at the firm.

The routine chores of daily home life also presented problems. He needed repeated reminders when shopping for newspapers and magazines, and even had to be prompted, on occasion, to do the shopping itself. This man of numbers and finance was having difficulty handling the sums needed to make incidental purchases at the local news store, but

Kumar, the owner, helped compensate for his confusion and errors. The weekly bridge game with the Kretchmers was now out of the question, but the couples continued to get together for coffee and dessert. Sally had to remind Enoch about upcoming appointments. She also had to help him recall recent outings, including the dinner last week with their daughter and son-in-law. While he easily recalled names of all family members, he did need help with names of acquaintances and even some close friends. Judith, the housekeeper, continually helped Enoch find his lost Mont Blanc pen, his keys, and his wallet. Without his devoted personal support system, Enoch would be adrift in an alien world of unrecognized landmarks.

Enoch had never been a daydreamer. Even as a child, he was always engaged in the social group. Whether small talk or weighty matters of the world, he was an active participant. He seemed to have had an opinion on most matters, whether it was the Yankees' pennant prospects, the Vietnam War, or the fare for dinner. Against this background, his present lapses were thrown into bold relief. Enoch drifted. Wednesday night, Pat and Gail Cowley visited for coffee, and were discussing an addition to their Sag Harbor retreat. They were adding a fitness room, and were exploring the virtues of different exercise machines. The debates moved from Cybex to Ubex to Tuff Stuff, names with which Enoch was intimately familiar.

The very moment before Pat turned to Enoch for advice, Sally's alarm system screamed. Enoch sat on the sofa, staring into space, totally oblivious to the world around him. He might as well have been in a coma. He might as well have been a thousand miles away. His eyes were open and betrayed emptiness. His face was expressionless. Beating Pat by a fraction of a second, Sally nearly shouted to Enoch that he needed more coffee. Enoch was convulsed back to reality, Gail spilled her coffee cup, and Pat was interrupted midquestion. The whole affair was a jarring embarrassment, but seemed to deflect attention from Enoch. Nevertheless, by the end of an evening of lapses and misremembered names, it was sadly evident to Pat and Gail that their close friend Enoch was having very serious problems.

Sally's anxiety, tension, crying spells, and anger were becoming overwhelming. She couldn't cope with the need for constant vigilance around Enoch. Whether alone together at home, shopping for groceries or carpets, or getting together with friends, Enoch required near-constant attention. She felt like a Seeing Eye dog. It wasn't just the picking up after, the post hoc damage control. Enoch's every awkward eventuality had to be anticipated.

Enoch's problems now began to directly affect Sally. It was only after an hour lying in bed at night, thinking of the day's events, that she realized that she couldn't sleep. Tossing somehow led to a fitful, half-awake state, but she was alert and anxious around 4:30 every morning and she dragged through the day exhausted. After lunch, she now needed coffee, a thoroughly distasteful habit, just to remain awake. And lunch itself was becoming a chore. She had always looked forward to soup and salad with friends. Her favorite had always been tuna and sun dried tomatoes on a croissant, but her appetite was a thing of the past. She remembered losing all taste for food years ago, with her bout of hepatitis. Then, she couldn't even look at food. This was different. Now she just didn't seem to care.

Preoccupied with Enoch, fearful of the future for the first time in her life, anxious about getting herself and Enoch through the day, Sally felt that she had lost all appetites. The morning news on the public radio station no longer held any attraction. She had no interest in her ritual of sitting on the Chesterfield couch with morning cup of coffee and newspaper with the amber rays of sun setting the airborne dust particles on fire. Just passing the time to lunch was a labor. Periodically, without warning, she burst into tears, which evaporated as precipitously as they appeared. She had no desire to see her friends anymore, much less have lunch with them. If emotions had sensations, then Sally had lost the smell, taste, touch, and feel of her experiences. Her vivid life of long ago had passed through a pastel phase and was now an indeterminate drab.

❖

The story of nerve growth factor, so far, has been the story of how nerve cells stay alive. We have barely touched the dark side of the story: Deprived of NGF, nerve cells die. Yet, to understand the actions of NGF deeply, cell death as well as life must be grasped. In fact, one of the revolutions in neuroscience and neurology has been insight into the nature of cell death. We now know that cell death occurring in a variety of disorders, including Alzheimer's disease, exhibits important commonalities. We return to stroke to illustrate.

Charlene Washington's sudden right-sided weakness and difficulty speaking were caused by a clot blocking a critical brain artery (Chapter 1). The clot prevented blood and life-giving oxygen from reaching specific brain centers. If blood and oxygen deprivation last long enough, nerve cells die. Only recently have we discovered that a second type of cell death, delayed death, also occurs in stroke, and in a number of other brain diseases.

The acute death of brain cells caused by the clot results in the release of cellular toxic signals that kill adjacent cells. The toxins elicit a second round of delayed cell death in stroke that is separate and distinct from the first round. The second round of death may be at least as devastating as the first. Delayed death is a feature of disorders as diverse as stroke, brain and spinal trauma, and Alzheimer's disease. Delayed death contributed mightily to the strange disconnection syndrome suffered by Charlene Washington.

Nerve growth factor deprivation causes the delayed type of death. This discovery led to deep insights into the nature of NGF actions, into the mechanisms by which NGF fosters survival, connectivity, and communication.

❖

Birth and growth are among life's happiest offerings, calling to mind first smiles, first steps, first words, and the sense of creative progress toward a mature perfected state. Many of us have long since repressed the sobering fairy tale lessons that happiness implies sadness, that progress is accompanied by regress, that bad is a constant companion to good. Yet we do realize, if only abstractly, that the living must make way for the unborn, that entire civilizations thrive and die, that our bodies grow by sloughing dead cells. The wheel of fate even deems that the once mighty Roman Empire rises and falls, that the invincible Spanish Armada meets defeat, and that the thousand-year Reich dies in decades. So too, brain development depends on death.

As we gain a clearer view of the astonishing role of massive cell death during a life, we appreciate the deepest secrets of brain function. Mechanisms in brain cell death are clues to the unprecedented complexity of brain architecture, which lies at the heart of mental function. The molecular events that govern brain cell survival and death continue throughout life. The function of surviving cells is altered by experience; they thereby form new and stronger synaptic connections, a key substrate of learning and memory. We must understand death to comprehend the life of the mind.

Death and Development

Though we associate creation, production, and construction with development, the role of death in development has been appreciated only recently. More than 80 percent of the cells produced in many brain systems are destined to die. How can we begin to understand such strange extravagance? How does life tolerate this apparent wastefulness? We may gain perspective by viewing cell death in the context of development in other species. By focusing on an extreme though familiar event that depends on death, we can appreciate its role in the economy of life.

Metamorphosis had been central to real-life development for millions of years before the first fairy maiden kissed a frog, magically transforming it into a prince. The metamorphosis of caterpillar to butterfly, or tadpole to frog is the inspiration for romantic metaphor from Bible to bard to Grimm. These conversions have epitomized creation. Yet death is central to this singularly developmental event. As tadpole becomes frog, its tail shrinks and disappears during the selective death of tail cells. The tail cells do not become something else; they die. When caterpillar retires to cocoon, massive cell death precedes the growth of cells that comprise the emergent butterfly. Without death, these transformations could not occur. The dramatic death of metamorphosis is completely typical of less cataclysmic events that occur during development, and throughout life. Additional examples make familiar the commonplace nature of death in development.

Cell death as a widespread if not universal phenomenon in development was recognized fifty years ago. An influential article was published in 1951 that reviewed the evidence for developmental cell death in most body systems.[1] The author of the then controversial article, A. Glucksmann, was an embryologist working (some thought appropriately) in Strangeways Laboratory in Cambridge, England. He compiled exhaustive evidence to prove that death was the rule, not the exception. He even classified cell death into types. One type, "Morphogenetic cell death," occurs during development and literally generates form. As the embryo grows from a solid sphere of cells to a hollow ball to a recognizable organism, highly selective cell death plays a critical role. Glucksmann also persuasively showed how cell death generates body form, from the heart to the blood vessels to the digestive tract to the nervous system.

Since function follows form, cell death assumed center stage in biological process. Cell death crafts the architecture of tissues and organs. Architecture determines how organs work. Architectural design reaches its zenith in brain and nervous system. Brain architecture defines mind architecture. To understand brain and mind, we had better understand the nature of cell death in the nervous system.

At the time of Glucksmann's article, death in the nervous system was being studied by none other than Rita Levi-Montalcini and her long-time collaborator, Viktor Hamburger, the great embryologist. Their landmark collaboration and its development is one of the inspirational epics of biology.

Rita Levi-Montalcini and Death

Levi-Montalcini graduated from medical school at the University of Turin in northern Italy in 1936.[2] Though she specialized in the clinical pursuits of neurology and psychiatry, she was also captured by pure science. Her role model was Professor Giuseppe Levi, for her "the master" who directed the Istituto di Anatomia in Turin with iron will and the highest intellectual standards. She was caught in the classic conflict. Should she become a practicing physician, or a scientist in the mold of her idol, Levi? While pondering her dilemma, fascist jackboots splintered the door. In June of 1938, Mussolini proclaimed the "Manifesto per la difesa della razza," which prevented all non-Aryan citizens from entering academic or professional careers. The world of the intellect had to be cleansed of non-Aryan contamination. To protect the integrity and purity of the superior race, marriage between Aryan Italians and Jews was forbidden. The dark shadow of anti-Semitism had moved south to Italy.

In 1939 Levi-Montalcini accepted an invitation from a neurological institute in Brussels. The respite was short-lived. Within months, Germany, England, and France declared war. As the invasion of Belgium rapidly approached, she rejoined her family in Turin, but life in Italy had deteriorated. To Rita Levi-Montalcini, Jew, all scientific laboratories were closed. Jews were barred from university libraries. Professions were forbidden. And the shadow now darkened all Europe: there were no safe havens. Her family refused to emigrate to the States.

Young Levi-Montalcini devised a characteristically unorthodox, daring course of action in the life-threatening Aryan world. She would secretly build a laboratory in her bedroom at home. Evincing the indomitable spirit that was to revolutionize biology and medicine, she recalls that she "was thrilled at the idea of this experience à la Robinson Crusoe." Though she had never performed a single experiment in embryology of the nervous system, though libraries of the Academy were closed to her, though she had no independent scientific funding, though she owned no scientific equipment, Levi-Montalcini plunged forward. It never occurred to her that it couldn't be done. She outfitted the laboratory with the bare minimum. She procured an incubator for tissue cultures; she had mastered culture techniques under the tutelage of Levi. She obtained a small microscope to examine embryos and tissues. She found a light she could use. She located an instrument to slice tissues for study under the microscope. She fashioned from sewing needles surgical instruments for ultra-delicate microscopic surgery and dissection. She shaped and sharpened the needles on a stone. Levi-Montalcini was nearly reinventing the scientific laboratory, not to mention the spirit of inquiry in a world gone mad.

Her laboratory was in a bedroom in Turin. The guiding light originated elsewhere. Denied the lifeblood libraries, her Bible and inspiration was an article by Viktor Hamburger dated 1934, which she had read years earlier.[3] Hamburger was studying how targets affect nerves that send fibers to the targets. How do the nerves know to select the correct target? How do the nerves find the right pathway to guide them to the targets? What effects do the targets have on the developing, ingrowing nerves? These questions are central to the growth of connections and the flow of information in the brain and nervous system. Hamburger used chick embryos to investigate these problems, and Levi-Montalcini adopted this experimental model.

Her small bedroom became a haven, a gathering place for Professor Levi and his adoring, former students. In a bit of a reversal, the professor, having lost his university position, asked to become Levi-Montalcini's assistant. Together, they carried out their experimental campaign, as the triumphant advance of the Germans all over Europe seemed to herald the end of Western civilization. They continued their experiments, while outside the daily drumbeat of virulent anti-Semitism reached a maniacal pitch. Finally, the Allied bombing of northern Italy became so intense that Levi-Montalcini and her family were forced to flee to a small country house. Professor Levi escaped to a tiny mountain village. Undaunted, Levi-Montalcini set up another laboratory, even more primitive than the first, under nearly impossible conditions. Chick embryos were in short supply. Electricity was disrupted every few days. Nevertheless, she completed a study of the development of hearing and balance centers in the brain. The article describing the work was published years later in 1949, in the United States—a triumph of the will.

But the night was growing darker. In the autumn of 1943, after Mussolini's fall, Italy was overrun by Nazis. Jews were hunted down and slaughtered in the streets; mass murder became routine; others were deported to concentration camps, gas chambers, and ovens. Levi-Montalcini and her family fled to Florence, losing themselves in a sea of refugees. Using a forged identity, she barely eluded detection and certain death. The fragile fabric of civilization was unraveling.

Somehow she survived until the end of the war in 1945. Levi-Montalcini returned to Turin and became Giusseppi Levi's assistant; he had been reappointed as professor of anatomy at the university. One year later, in 1946, Levi-Montalcini received a letter from the States. It was from Viktor Hamburger, one of the founders of neuroembryology. He had read a 1942 scientific article by Levi-Montalcini that postulated a novel interaction of targets and nerves, a view which differed from his thinking. (The Jewish authors, Levi-Montalcini and Levi could not publish the article in a journal in fascist Italy. Rather, it was published in *Archives de Biologie* in Belgium.[4]) Hamburger was fascinated by it. He invited Levi-Montalcini to join his laboratory at Washington University in St. Louis for one to two years to collaborate on the problem. In 1947, she left Turin for St. Louis, where she spent the next twenty-five years.

Life and Death in the States

In the States, Levi-Montalcini reveled in the twin miracles of Hamburger and the American university, embodiments of a European freedom long vanished. Viktor Hamburger was the incarnation of one of the great traditions in biology. In Germany, he had been a student of the renowned embryologist, Hans Spemann. Hamburger was Spemann's designated favorite son, the bearer of the torch in the next generation. Spemann received the Nobel Prize in 1934 for his discovery of induction in the embryo: one group of cells acts on other populations to induce new tissues and organs. Cells signal each other in the embryo and spur the development of new tissues. Through signaling induction, the embryo bootstraps itself through development. As heir to the Spemann tradition, Hamburger turned his attention to embryology of the nervous system. He was pioneering a study of how targets and their nerves interact during development.

In 1932, as Germany was slipping into Hitler's dark night, Hamburger visited Chicago to work for a year in the laboratory of Frank Lillie, a friend and colleague of Spemann. Lillie had introduced the field of embryology to the use of the chick embryo, which yielded a wealth of answers to the riddles of development. With the victory of Hitler in 1933, Hamburger turned away from his homeland and remained in the States to keep learning about the developing nervous system. In 1935 he accepted the chairmanship of the Department of Zoology at Washington University, and continued the work he had begun in Germany. Like Einstein, Fermi, Teller, and Szilard in physics, with Hamburger the continent of Europe was losing a shining star, this time in biology. America employed its genius for welcoming European talent and human capital.

For Levi-Montalcini, the contrast between the modest, self-effacing, quiet Hamburger, and the dictatorial, iron-disciplined, booming Levi epitomized the European–American dichotomy. Hamburger, ever kind and considerate of others' sensibilities, would signal disapproval with the gentlest of remarks. Levi's gentlest rebuke, recalled Levi-Montalcini, was a courtly "I beg your pardon, but you are a perfect imbecile."[2] For Levi-Montalcini, the European visitor scarred by suffering and death of the Great War, the open, American university was a garden of Eden, a paradise. Levi-Montalcini's intuition, passion, and visceral insight were perfect complements to Hamburger's thoughtful, incisive clarity. The collaboration changed the direction of developmental biology. They knew that targets and nerves interact in some fashion; the key was to discover the mediating signals.

Levi-Montalcini and Hamburger embarked on landmark experiments with the chick embryo. They examined the nerves that transmit sensation and discovered that the nerve cells are lined up in groups along the developing spinal cord from neck to tail. The nerves send fibers throughout the body and communicate sensations back to the spinal cord. During normal development, many normal nerve cells died and degenerated. Death was not random. Detailed inspection revealed that the cell death followed a pattern in space and time.[5]

Death was overwhelming in the neck and chest regions. In dramatic contrast, the number of degenerating cells was far fewer among nerves sending fibers to, or innervating, the developing wings and legs. The large target areas of the wings and legs were associated with the more robust survival of innervating sensory neurons. Ingenious experiments supported the idea that target size had something to do with nerve death.

Delicate surgical removal of the developing wing in the embryo markedly increased the death of neurons destined to innervate the wing. Removal of a developing leg similarly prompted the death of nerves that were to send fibers to the leg. Reducing target size increased naturally occurring neuron death. Target size appeared to determine how many nerve cells lived and how many died.

If the removal of a target killed more developing neurons, what would happen if the target size were made larger? To examine this, Levi-Montalcini and Hamburger transplanted an extra limb to the developing embryo. They then carefully examined the regional sensory nerves under a microscope. Their observations were astonishing and compelling: Neuron death was profoundly diminished in the region of the extra limb. Greater target size rescued neurons from death.[5]

These revolutionary experiments led to radical insights into how the nervous system is built. Massive nerve cell death is, most generally, a normal part of development. Death crafts the form of the nervous system. More specifically, target size determines how many neurons live and how many die—an efficient mechanism for matching nerve cell number to the needs of targets. It explains Levi-Montalcini and Hamburger's early experiments. Cell death is more frequent in the neck region, because the neck is small, and in the chest, where only the chest wall receives nerve fibers. By contrast, the wings and legs are huge outgrowths that receive innumerable nerve fibers. In these areas far more nerves survive.

Cell death, then, is a prime architect of the developing nervous system. Normal developmental neuron death contributes to neural architecture, which generates function and ultimately mind architecture. Can we generalize from these pioneering studies performed on sensory neurons? Does cell death occur in other neural systems? Is this startling phenomenon widespread? The answers are resoundingly affirmative.

Levi-Montalcini herself extended these studies to neurons in the spinal cord and found that developmental neuron death occurred there as well. Hamburger subsequently examined the motor neurons that control muscle movement and discovered that they, too, undergo massive death during development. In sum, Hamburger and Levi-Montalcini demonstrated that developmental death was widespread in the nervous system. Scientists following in their footsteps discovered neuron death in areas throughout the brain. Just as important, the new experiments revealed how cell death sculpts the brain to account for architectural details that may be decisive in mental function.

One particularly illustrative example is a group of brain neurons that sends fibers to the retina in the developing eye.[6] When the group first forms in the embryonic brain, it is roughly spherical and lacks any distinctive structure. At this point it contains about 27,000 cells. Over the next few days this ill-defined cell group transforms into an intricately patterned mature structure due to a precisely synchronized wave of neuronal death. The cell number is dramatically reduced to 11,000. During this period, cells die at the rate of 200 to 300 per hour! Recapping the Hamburger–Levi-Montalcini experiments, removal of the eye markedly increases cell death. Indeed, with eye removal the cells disappear entirely. During normal development, the precisely orchestrated pattern of cell death converts the primitive sphere of cells into an intricate, folded sheet. The sheet forms a precise map of the retina. Any point in the group of cells corresponds exactly to a point in the retina. Any point on the retina corresponds, in turn, to a point in visual space. In this instance, cell death forms a map corresponding to the outside world; cell death results in precise point-to-point communication in the brain. Cell death generates an architecture that directly underlies visual perception.

A flood of subsequent experiments indicated that developmental cell death is virtually ubiquitous in the brain. Death crafts areas that govern sensation, perception, muscle movement, cognition, and emotion. Death is not simply a necessary waste removal exercise, cleaning up useless scraps of material at the construction site. Death is central to building the architecture from which mind and emotion arise. While counterintuitive, the reality is that death builds brain and mind. Theoretical studies also emphasize the importance of cell death in artificial intelligence.

Theoretical Advantages of Death

A fruitful exploration of how brain and mind work developed a new field that emerged in the 1980s: computer-based neural networks. The new area arose from a synthesis of psychology, neuroscience, computer science, and so-called artificial intelligence. The underlying idea is that knowledge exists in the connections between neurons. Learning, in this view, consists of modification of the strength of connections among neurons. So, the field has also been called "connectionism." Neural networks, composed of artificial neurons modeled in computers, exhibit startling feats of learning and intelligence and hence resemble live brains.

Neural nets can learn how to speak, how to recognize objects, and how to solve new problems. Surprisingly, the nets have shed light on the seemingly unrelated issue of neuron death.

In practice, a computer-based network is presented with a problem and asked to solve it. By analyzing the steps taken by the neural network in problem solving, the rules governing thought are, one hopes, uncovered. In these experiments, death suddenly became relevant. Excessive numbers of artificial neurons impede the process of *generalization* by the network; that is, the network had difficulty generalizing that cats have four legs from a group of sample cats, or picking out the common features of trucks, if the network had too many neurons. There was more.

Networks are better at solving difficult problems if they start with a large number of artificial neurons and then reduce them to a smaller number. Networks that start with a smaller number of neurons are just not as intelligent. The loss of artificial neurons helps learning. Artificial death of artifical neurons helps artificial intelligence. How in the world did computer neuroscientists come up with this strange conclusion?

In the first sentence of a seminal article, computer mathematicians Michael Mozer and Paul Smolensky of the University of Colorado presented a decidedly nonscientific introduction that captured a common feeling.[7] To wit, "One thing connectionist networks have in common with brains is that if you open them up and peer inside, all you can see is a big pile of goo. Internal organization is obscured by the sheer number of units and connections." They proceeded to ask "What is the importance of each neuron in a computer or brain in solving a problem or accomplishing a task?" They wanted to discover mathematical criteria to define the relevance of a neuron in task performance. There were compelling reasons for wanting to eliminate the least relevant neurons and thereby simplify the network. It wasn't simply a matter of getting rid of the goo.

First, by eliminating neurons that serve little or no purpose, the sheer number of issues that a network has to consider is reduced. This limits the range of generalizations (or solutions) for a problem, and limits incorrect solutions.

Second is the matter of speeding up learning, which is fast when there are many neurons. Yet a large number of neurons simply allows the network to arrive at too many solutions, or generalizations, many of which are close but incorrect. Learning is slower with fewer neurons, but generalization is better. Their novel idea for the best of all possible network worlds was to teach the network with many neurons, and then eliminate the irrelevant ones. They predicted that this procedure would prompt rapid learning, and then gradually improve generalization performance, as neurons were eliminated.

They proceeded to derive complex equations to achieve their goal. They found a measure to determine whether a neuron serves a critical function in a network, which they determined by asking mathematically what happens to the network's performance when a unit (neuron) is removed? Their approach worked; they identified and removed irrelevant neurons. They analyzed neural network performance with and without neuron elimination (read "cell death"). Their discoveries confirmed their predictions.

Using their mathematical approach, they identified "rules" of learning that dictate the important role of cell death. Neurons critical for solving a problem could be identified, and so could neurons that conveyed redundant information. They distinguished neurons responsible for correctly solving most problems from ones that only dealt with exceptional cases. As a happy byproduct of their enterprise, they discovered the minimal features from the environment that a network needed to identify commonalities, and thus generalize about cats or trucks.

These rules elucidated the critical functions of cell death. For certain problems, static networks were unable to discover solutions, while networks simplified through cell death never failed. Moreover, the network subjected to neuron elimination learned faster. For other problems, networks that experienced neuron death required less teaching.

In summary, a network used spare neurons to learn rapidly. As neurons were eliminated, the network discovered a more concise characterization of the regularities of the task. The process of neuron elimination also allowed the network to avoid solutions to a problem that were close, but incorrect.

These startling discoveries suggest that cell death in the living brain is key to the development of mind. In a way that we do not fully understand, neuron death itself is central to the genesis of intelligence. It became critical to discover the cellular processes underlying neuron death.

How Neurons Die

How do neurons die? Well, first we should classify the types of death. The all-too-familiar death of individuals could be used as a rough road

map; it enables us to think of the different deaths that apply to brain neurons. The human species has been witness to bewildering styles of death.
As a group endowed with singular ingenuity, we have even invented
novel forms of death. Are any of these relevant to brain cell death? To
orient this funereal undertaking, here are some general classification
schemes: death may be natural (old age) or premature, it may be elicited
by external causes or internal malfunction, it might result from individual
(heart attack) or social (as in war) factors, it could be traumatic or
nontraumatic, it might be caused by deprivation (as in starvation or vitamin deficiency) or gluttony (leading to obesity, atherosclerosis, heart attack, or stroke), it could be generated by genetic or environmental
circumstances. The schemes could go on forever. Let's choose a sample
classification scheme arbitrarily and pursue it systematically to learn more
about the deaths available to willing brain neurons.

Honoring our Western biomedical tradition, we adopt a conventional approach. Death can be attributable to internal defects or external
agents of one sort or another. This classification is the dark side of nature
versus nurture. Beginning with the latter, we identify external agents and
lethal events. Environmental influences from arsenic poisoning to auto
accidents to bacteria that cause pneumonia or meningitis are examples of
death by nurture. By analogy, neurons could die from toxins, trauma, or
infection.

Internal causes of death include genetic ones, as with the hemophil-
iac who hemorrhages in the absence of treatment, the Huntington's patient, or the centenarian who dies from old age. At the extremes we can
distinguish internal versus external causes of death and ask how either is
relevant to brain neuron death.

Our little scheme can encounter difficulty in gray areas. Where do
we classify the unique human invention of suicide? Is suicide external or
internal, nature or nurture? That could be fodder for an interesting af-
ter-hours debate; for now let's stick to brain neurons and ask whether
death in the brain during development is nature or nurture, internal or
external.

How to approach this question? Like any competent physician, we
would have to catalogue the signs and symptoms of developmental cell
death. With this clinical profile, a diagnosis of developmental death could
be made with confidence. We could avoid confusing developmental
death with other cellular diseases that are irrelevant. As it happens, developmental biologists examined this issue twenty years ago.

The Discovery of Cell Suicide

Four decades ago, Glucksmann had documented the occurrence of developmental cell death in every system in the body, as we have described. Twenty years later, several prototypical cellular symptoms were identified. John Kerr, at the University of Queensland School of Medicine in Australia and Andrew Wyllie at University Medical School in Edinburgh, characterized the symptoms of cells suffering developmental death.[8] They shriveled up and shrank in size, like a grape turning into a raisin. The cell membrane developed blisters that Kerr and Wyllie called "blebs." The complex of DNA and proteins in the nucleus, termed "chromatin," condensed, and the DNA underwent a characteristic type of breakdown. This triad of cell size, membrane, and DNA symptoms distinguished developmental death from all other forms of cell death. This characterization enabled one to view developmental cell death as an unambiguous, discrete cellular malady.

The developmental cell death syndrome was so unique that Kerr and Wyllie felt constrained to honor the hallowed medical tradition of inventing a new name for a new disease. Not the least of their contributions was picking a name that every scientist seems to pronounce differently: "apoptosis," a term derived from the Greek for a tree losing its leaves.[8] The cellular changes of this form of death were so stereotyped that it was also called "programmed cell death." But there was more.

This characteristic cell death was not restricted to development. The Apoptosis Grim Reaper claimed diverse cells throughout life, under different circumstances. For example, in the immune system, which guards the body against foreign invaders, cells depend on growth factor signals. These life-supporting signals are analogous to growth factors in the brain. Immune cells deprived of these "cytokines," which are cellular hormones, undergo apoptosis. Biologists discovered other instances of apoptosis: a variety of cells depend on endocrine system hormones; and prostate cells undergo apoptosis when deprived of testosterone, the male sex hormone. Hence, changes in the environment cause a peculiar, stereotyped cell death. What is the cellular disease, and how does it affect neurons?

The mystery attracted the attention of Eugene Johnson at Washington University. In the extraordinary tradition of Hamburger and Levi-Montalcini, Johnson had already made central discoveries in the world of NGF. Johnson always came up with offbeat approaches to thorny scientific problems. His novel experiments may have reflected his unorthodox entry into science. Johnson began his working life as a druggist. In an August world of Ph.D.s, M.D.s and M.D.-Ph.D.s, filling prescriptions, pill-pushing, and peddling vitamin supplements did not necessarily constitute a shortcut to the frontiers of science. Nevertheless, he gradually moved from the pharmacy to teaching pharmaceutical science to laboratory research. As the romance of the drugstore faded, he joined the faculty at the Medical College of Pennsylvania, the original Women's Medical School in the United States, and followed his intuitive nose into the scientific laboratory. His unorthodox background may have been a blessing. Johnson was unencumbered by the intellectual rules and regulations that guided, and severely shackled, many scientists. He simply did not appreciate that certain experiments could not be done. So, he did them—successfully.

Johnson was obliquely following the spirit of advice that Julius Axelrod, former lab helper cum Nobel Prize winner (Chapter 3), offered to his students. Axelrod, who also had entered science through the backdoor, counseled students about to enter a new scientific area *not to read the scientific literature*. According to Axelrod, the oracle, the literature only tells you what cannot be done. Minds are closed before the first shot is fired. Axelrod advised disciples to follow their noses to the improbable hypothesis, unlikely experiment, and unusual result. Above all, stop talking, and do the experiment. Axelrod's scientific offspring have long since revolutionized neuroscience, pharmacology, biochemistry, and therapeutics.

In the 1980s, most scientists in the field were studying how NGF supports neuron survival, how it saves life. Johnson, by contrast, wanted to know how NGF deprivation kills.[9] Though the question was unorthodox, Johnson used a standard preparation. He cultured sympathetic neurons, keys in the fight-or-flight system. Sympathetic neurons require NGF in the bathing medium to survive. Johnson's idea was to remove NGF from the medium and observe the neurons to assess how NGF deprivation results in death. In the presence of NGF, the neurons thrived beautifully, as expected, growing in size and sending out characteristic processes. Johnson then removed NGF and waited. The neurons showed no changes under the microscope for the first few hours. Within twenty-four hours, however, disaster struck. The fibers that each neuron had sent out began to fragment and the neurons shrank. Blebs appeared on the neuronal membranes and the nuclear DNA underwent degradation. The neurons suffered typical apoptotic death.

Johnson and his student David Martin performed a crucial experiment.[10] They wanted to know exactly how NGF deprivation, like a form of starvation, caused the neurons to progressively waste away. Alternatively, NGF removal could trigger pathological processes. Johnson and Martin approached things in an unusual fashion. After removing NGF, they treated the cells with drugs that shut down all protein synthesis. Proteins are the workhorses of all cells and form the building blocks for life and growth; their enzymes form the machinery that processes all cellular material for the energy, metabolism, and growth necessary for life. Johnson and Martin obtained a result so strange that it changed our thinking about cell death. Blocking protein synthesis saved cells from the death resulting from NGF deprivation.[10]

Protein synthesis, a hallmark of life, was necessary for neurons to die! The neurons were not simply running down after NGF was removed; the removal was not a form of starvation in which systems simply got used up and failed. Death depended on the production of proteins. The conclusion seemed unavoidable: when NGF was withdrawn, the neuron synthesized killer proteins. This was the first indication that cells committed suicide when NGF was eliminated. Apoptosis or programmed cell death might be better termed "programmed cell suicide." NGF removal prompted a neuron to turn on an internal suicide program.

Where in the cell did the suicide program reside? Because proteins are synthesized by genes, it was not a great leap to begin searching for the killer genes. While experiments in the formidably complex nervous systems of vertebrates, such as mice and rats, were difficult and initially unrevealing, help came from other quarters.

Scientists working with a very simple invertebrate animal, the nematode or roundworm, provided the breakthrough. Sydney Brenner, one of the original pioneers of molecular biology (see Chapter 1), latched on to this animal in the late 1960s to study a nervous system devoid of the dizzying complexity observed in vertebrates. In contrast to the billions of neurons in vertebrates, the roundworm had only about 1000 neurons. Each individual neuron could be identified and was identical from roundworm to roundworm. As important, neurons in the roundworm also underwent programmed cell death. The stage was set.

In the early 1980s Robert Horvitz's team at the Massachusetts Institute of Technology identified two death genes in the roundworm.[11] Protein products of these genes caused developmental neuronal death. The genes were named ced-3 and ced-4 (as in cell death). Much to their surprise, they discovered another gene, ced-9, which prevented cell death. A cell that activated ced-9 was rescued. Does this strange genetic yin/yang of mortality in the lowly roundworm have anything to do with life and death in the vertebrate nervous system?

The vertebrate immune system provided clues. An oncogene (cancer-producing gene) was identified in a form of leukemia that affects lymphocytes. The oncogene, named bcl-2,[12] is activated in B cell (lymphocyte) leukemia and prevents cell death, leading to this threatening disease. By inserting bcl-2 into cultured neurons, death is prevented upon NGF withdrawal. Moreover, bcl-2 is normally expressed throughout the brain of mice at the very time that neuron death is normally occurring during development. This discovery suggested that bcl-2 plays a role in the selective survival of specific neurons during the period of massive developmental cell death. Finally, bcl-2 bears a family resemblance to the ced-9 gene, which prevents death in the roundworm. Scientists cautiously concluded that programmed cell death is widespread in the animal kingdom, and that molecular mechanisms can be shared by distantly related animals. Worms, then, provide insights to human death.

Building on this earlier work, many other experiments on apoptosis have provided a glimpse of the cell's internal world: the forces of life and death are on hair-trigger alert. NGF and other growth factors activate survival genes such as the bcl-2 family, which ensure survival. Growth factor deprivation activates suicide genes with names like bax, c-myc, and p53. Thus the neuron is hardwired for both survival and suicide. External signals determine whether the survival program or the suicide program is turned on. Life and death in the nervous system is not a casual affair.

As we learn more about the intricacies of life and death in the brain, we realize that programmed cell suicide is not just about development or normal aging. Apoptosis plays a critical role in many brain diseases. Stroke, head trauma, spinal cord injury, and even Alzheimer's disease are due, at least in part, to the pathologic activation of the apoptosis program. Neuroscience moves rapidly from the abstractions of building brain maps to life and death at the bedside.

On the Origins of Cell Suicide

The forces of self-destruction are on call in the cell, waiting to be activated. But this is a strange way for life to behave. How in the world did this improbable scenario evolve? What did nature have in mind by making cell survival dependent on the vigilant repression of genetic suicide programs? What is the conceivable advantage of building systems in which cell life depends on signals from other cells, growth, and trophic factors, to suppress suicide? We can imagine that somehow cells in multicellular organisms commit suicide because they are damaged, redundant, toxic to others, or maybe even cancerous, thereby favoring the organism's survival. But cell suicide is far more ancient, having appeared in single-celled life forms.

Programmed cell death occurs in single-celled organisms that originated 1 to 2 billion years ago. These forms of life far preceded the evolution of multicellular plants and animals, which appeared about 0.7 billion years ago. Slime molds, the single-celled trypanosome (which causes the human disease, trypanosomiasis), and the free-living, rapid swimmer, tetrahymena, all commit suicide. Their self-destruction is triggered by environmental conditions, as in multicellular forms. Environmental stress, that is, conditions that prevent growth and differentiation, precipitate suicide. Extracellular signals, then, regulate life and death in these primitive forms, as in the brain.

What is the selective advantage of programmed cell death in the evolutionary arms race? Though we don't know for sure, there has been no shortage of fascinating speculation. Single-celled organisms frequently aggregate in colonies, forming transient multicellular structures that mimic true multicellular organisms. Programmed cell death may encourage unrelenting selection for the fittest individuals in these colonies of unicells. The selective survival or destruction of cells may enable the colony as a whole to adapt to environmental conditions by adjusting the number of cells, the configuration of the colony, and the coordination of cell division within it. Thus programmed death may play a morphogenetic role in these colonies, just as Glucksmann demonstrated for true multicells.

Selection for killer genes may follow several scenarios. Competition among individuals in colonies of different species may have favored genes encoding toxins used to attack foreigners. Simultaneously, however, genes protecting against the toxins would have conferred selective advantage, and the beginnings of survival genes may have made their appearance. Speculation has ranged from the probable to "just so" stories. There is a categorical alternative to the interindividual competition hypothesis.

In the other general hypothesis, programmed cell death is closely tied to cell division and reproduction itself, and to the genesis of genetic diversity. It is a solution to the unavoidable, sometimes devastating, problems that arise as a cell goes about the business of life. A sense of such problems and solutions may be gained by considering mutation. The cell is unable to completely avoid random genetic mutation. None of us, including our cells, is perfect. In the reproduction of the DNA code, errors, no matter how rare, will occur. The errors can be seriously debilitating, if not fatal. To combat them, cells evolved intricate mechanisms to proof-read DNA, detect errors, and eliminate them. Nevertheless, the errors are a natural and unavoidable consequence of life.

Life requires mitosis, the complex coordination of DNA synthesis, doubling of the genes, separation of chromosomes, cell division (involving multiple synchronized steps), and coordinated growth and differentiation of the daughter cells. This is the "cell cycle." One misstep, one error of timing can wreak havoc and lead to a monster cell. It so happens that many genes that control the cell cycle also control programmed cell death. In this view, programmed cell death may have originated as a natural consequence of errors in the cell cycle, which decrease fitness and result in apoptosis due to action of these cell cycle/apoptosis genes. Natural selection governs the cycle precisely to suppress suicide. Failure results in the death of cells that are less fit.

Cell Death, Brain Maps, and Reality

Though the story is far from complete, we can trace the dim biological trail of cell death from single-celled organisms to brains and minds. Apoptosis arose in unicellular organisms as an unavoidable consequence of life itself, whether through interindividual competition or as a byproduct of the cell division cycle. Once established, cell death came to occupy a central role in the strategy of life. With the evolution of multicellular animals, and complex organ systems, apoptosis became pivotal. Selective, patterned cell death during development was co-opted to generate the complex forms, the architecture, of organs and systems of organs on which function is based. The hand develops through the selective death of cells that make up the primitive webbing connecting the maturing fingers in the embryo. The heart develops its complex, four-chambered mammalian structure through selective death and folding of the primitive cardiac tube. Architectural complexity reaches its zenith in the brain.

One key feature of brain architecture is the organization of maps that represent the external and internal worlds. This is readily appreciated by considering sensory maps. For example, regions of the cortex contain maps of the body plan. Touch or pain in any body region results in the firing of brain neurons devoted to that region. We perceive the stimulation of a body part through this selective action of neurons. Thus activation of parts of the map by experiences leads to our sensory perceptions, and this activation is relayed to the brain's motor parts that plan and execute movement. Map structure leads to map function, which forms our reality. Maps define our actions and reactions to that reality. Maps and their properties *are* our reality. We do not respond to all intensities of touch, for example. We can only sense a tiny band of energies in the entire electromagnetic spectrum and are oblivious to cosmic rays, x-rays, and radio waves. We hear only a tiny fraction of the audible energy waves in the world. The poverty of our odorant (smelling, olfactory) maps all but exclude us from the world of odor. The wide world of scents is a salient, life-and-death reality for most mammals, including our canine and feline companions; but it is almost nonexistent for us, not part of our reality, our brains, and our minds.

We are endowed with visual maps of considerable size and complexity, making our visual world rich by contrast. Yet this sense too is shaped and limited by the perceiving brain, and perceives reality no more objectively than any other sensory modality, whether olfaction or audition. The organization and operation of our optic system and visual maps endow us with the stereoscopic vision that renders our world three-dimensional. Separation of the plate from the table, of the tiger from the tree, recognition of predator, distinction of focused foreground from fuzzy background, detection of baseball movement at 60 mph toward a moving glove, calling an imaginary scene to mind—all depend on brain architecture crafted by cell death. Our reality is truly in our brains, not out there. Programmed cell death builds the architecture that generates the mind that is our reality.

Apoptosis and Brain Disease

Developmental cell death has been recognized for nearly half a century. Only within the last few years have we realized that cell suicide, a disaster waiting to happen, plays a key role in brain disease. Apoptosis plays a prominent role in disorders from Alzheimer's to trauma to stroke. The unraveling mystery illustrates how insights into a basic brain mechanism, such as programmed cell death, can lead to new treatments for life-threatening diseases.

Though the details are sketchy, the contribution of cell suicide to the tragedy of stroke is beginning to emerge. Stroke, along with heart disease and cancer, is one of the leading causes of death and disability in postindustrial society. Paralysis, loss of speech, blindness, and dementia are only some of the more dramatic deficits that compromise the lives of survivors. Yet treatments have been nonexistent. Available therapy has consisted of administering anticoagulants to prevent further blood clotting in the vessels supplying the brain. TPA and related drugs have been used to attempt to bust the clots that have already formed. The objective has been to restore circulation through clogged arteries. No treatments had been devised to directly prevent neurons from dying, or to rescue sick neurons. Even contemplating such an approach required understanding how neurons die in stroke.

The death of neurons, which causes the devastation in stroke, occurs in two stages. With the initial blockade of blood flow, neurons die due to blood and oxygen deprivation. This so-called necrotic cell death has been known for decades, and is similar to death resulting from a variety of insults. The big surprise, and cause for hope, came with the discovery that the second wave of death is due to apoptosis.[13] And frequently, the most devastating brain damage is due to this second wave of neuron death. While our understanding of exactly how programmed cell death occurs in stroke remains fragmentary, the outlines are emerging. The initial blood clot, which radically reduces or blocks blood flow, damages neurons in the area, prompting them to release neurotransmitter signals that flood that area of the brain. Normally these transmitter signals are released at fleetingly low levels, but damaged neurons release toxic levels of transmitters. The transmitter signals that cause damage in stroke are excitatory. They normally cause neurons to fire impulses. At the high levels released in stroke, however, these transmitters become "exicitotoxic," exciting neurons to death. The overexcited neurons die a programmed cell death. The flood of supranormal levels of excitatory transmitters in the brain triggers massive programmed cell death, a major culprit in the devastating results of stroke.

New discoveries are pointing the way to entirely novel treatments for stroke. Treatments are being designed to prevent the release of excitotoxic transmitters, block their action by blocking their receptors on neurons, prevent the initial events through which they trigger apoptosis, prevent the action of the killer genes, and prevent the effects of the killer proteins that the genes synthesize. Stroke, the hopeless curse that left cripples in its wake, is now a potentially treatable disorder. So too for degenerative neurologic disease, including Alzheimer's, and for brain trauma; they are yielding to an understanding of the nature of neuronal death.

Having endured brain death, let's consider the happier topic of birth and the brain, the subject of the next chapter.

❖

Enoch is experiencing increasing difficulty with the chores of daily living, emerging from his morning bathroom ritual half shaven, still partly lathered and unkempt, and leaving dinner food-stained and disheveled. On one occasion, he temporarily forgot how to brush his teeth and on another how shave. He quickly recovered. Crossing a threshold, he awoke one morning soaked with the smell of urine. If only dead brain neurons could be replaced. But a century of study led to the dogma that mammalian neurons do not divide after development: among all cells, neurons are not renewable resources. Stroke, trauma, and Alzheimer's lead to crippling, irreparable damage. Nevertheless, as we enter the twenty-first century, a scientific revolution is indicating that new brain neurons can be produced after all. Our concepts of brain function, and of treatment are undergoing profound reevaluation.

❖

6

BRAIN BIRTHDAYS

❖

ENOCH'S ABSENT-MINDEDNESS took an ominous turn over the ensuing six months. His seemingly innocuous difficulties with the routine negotiation of social engagements and mundane chores of daily life gradually transformed into episodic personal indifference. With greater frequency he emerged from the bathroom in the morning with hair not fully combed and his shirt was not completely tucked into his pants. On several occasions he missed loops on his stylish belts; and on a few suspender days he ignored a button, strikingly uncharacteristic of Enoch Wallace, Esq. Sally's sympathetic comment, "Enoch, your shirt," usually elicited corrective action, but at times she was greeted by bemused puzzlement. There was rarely a verbal response.

Dining became a new experience. Over the same period, no longer was Enoch toastmaster of the table, guiding gourmand, and enologist extraordinaire. Nowadays he often finished a meal with crumbs unaccountably on his lips and cheeks, and stains on his shirt and tie.

Of all things, Enoch even experienced changes in his morning bathroom ritual. He always began the day with a cold, bracing face wash. He continued that as usual with the same refreshing, optimistic sense that he always experienced. Once alert, he customarily brushed his teeth vigorously, using his Reach toothbrush, and the Crest gel toothpaste that he adopted several years ago. The scrubbed face and teeth always resulted in a sense of well-being that seemed to set the day right. This morning, he held the tube of toothpaste in his left hand, with the brush in his right hand, and, as a stranger, watched himself squeeze the toothpaste on the handle of the brush. He stared at the toothbrush, sensing something was wrong, but not knowing exactly what to do. Then he quickly rinsed brush and paste and started again. This time, he unconsciously spread the toothpaste on the bristles, as usual, and proceeded to brush his teeth without event. He gave the strange little episode no further thought.

Several days later, he had a momentary lapse while shaving. With shaving cream on his face and a double-edged safety razor in his right hand he was about to begin the procedure that he had watched in the mirror thousands of times. Enoch always began by smoothly stroking down from sideburn to jaw angle over the right cheek. Enoch looked at the razor in his right hand, looked at the face in the mirror, but was not sure what to do next. He rinsed the razor in the pool of water in the sink, staring into his face in the mirror, seeking an answer to questions he couldn't articulate. He placed the razor on the sink, and paused for a moment. Suddenly, he picked up the razor in his normal fashion and shaved as he had for decades. At least, almost as he had for decades. He emerged from the bathroom with the shaving cream not completely removed. He appeared at breakfast having missed small areas on his upper lip and chin. Fortunately, Sally was able to remedy the minor oversights.

Though Enoch did his best to minimize his puzzling episodes, though he attributed his lapses to an unaccustomed absent-mindedness, there was, of course, something more. He was experiencing bits and pieces of a strange abnormality known to neurologists as "apraxia." In this bizarre syndrome, pure motor-muscle function is normal. However, the ability to perform routine, stereotyped motor acts is deranged. Waving goodbye, lighting a match when presented with a matchbook, using a knife and fork become fragmented and difficult to perform. The underlying brain physiology is poorly understood, but serious pathology is almost always associated.

One day in June he became terrified by an alien episode. He awoke on a crystalline Saturday morning in eager anticipation of a trip to East Hampton to visit his daughter and son-in-law. The Wallaces had been planning the weekend for months. As Enoch emerged from his fragmented dream, still only partly conscious, something undefined was different in the world, bedroom, or bed. With sickening clarity he realized that his pajama bottoms were wet. The odor was unmistakable. Enoch had wet his bed during the night. A threshold was crossed.

❖

The slowed movement, muscle rigidity, crippled walking and tremor of Jeanette McCready's Parkinson's disease responded to the miracle of L-DOPA therapy (Chapter 3). Treatment transformed Jeanette from a near-wheelchair victim to a functional, hard worker once more. Thirty years ago, the very idea that brain disease could be treated was heretical, revolutionary. Reports of successful L-DOPA therapy were initially greeted by the medical community with thorough skepticism, as the fantasies of fakirs and charlatans. Yet for all the ultimate success of this new approach, treatment was only symptomatic: it did not address the underlying cause of Parkinson's disease. Although L-DOPA could buy seven to ten years of fruitful, productive life, Parkinson's patients inexorably progressed to a bed-and-wheelchair existence.

L-DOPA replaced the missing dopamine brain (transmitter) signal, alleviating Jeanette's symptoms. L-DOPA did not alter the basic, underlying disease process, the ongoing death over years of a critical group of neurons at the base of the brain. L-DOPA provides increased signal to the ever-diminishing pool of remaining neurons, overriding the symptoms early in the disease. Ultimately, as more and more neurons die, even L-DOPA becomes ineffective.

Though we don't know the cause of progressive neuronal death in Parkinson's, transplantation of replacement cells into the brain represents a promising new therapeutic approach. Rats, mice, or monkeys with experimental Parkinson's experience remarkable recovery after transplantation of neurons that send the dopamine signal to the correct brain target. While not yet perfected for patients such as Jeanette, the direction is clear. The implications are equally clear. If we could only induce the patient's own dopamine neurons to divide, to reproduce, we could replace the dying neurons and prevent, cure, or at least, abort the disease.

As every biology student knows, however, neurons do not repro-duce after development. This dogma, too, is in the midst of a profound revolution, as we shall now discuss.

❖

If you accept the counterintuitive notion that death is central to brain and mind function, you will be hardly surprised to learn that birth is also rele-vant. Yet the place of cell birth in brain function remains a deep mys-tery—one shrouded in hopelessness.

The dogma has been quite simple: Brain cells do not regenerate. Recovery of function after brain illness and injury does not occur. In the magnificent march of medical advance from shamans to the barber-sur-geons of the Middle Ages, to the antiseptic operating rooms of today, the brain has remained a nonrenewable resource. The tragic, maddening im-mutability of brain neurons is remarkable in the face of novel discoveries and treatments that have transformed every other sphere of medicine. Yet some notable breakthroughs place the sorry state of brain function and dysfunction in perspective.

Broken bones are routinely reset and heal normally; permanent crip-ples have, sadly, simply lacked access to basic medical care. Deformity and death due to malnutrition and vitamin deficiency are understood and pre-ventable, not visitations of demons. The horror of a fully conscious child held down by parents while enduring an amputation has been eliminated by anesthesia. Scourges of mankind, including pneumococcal pneumonia, tuberculosis, meningitis, malaria, and a plague of other killers are treatable with antibiotics. Vaccines have eliminated the threat of smallpox, diphthe-ria, and polio from entire societies. In the face of comparable relief from scores of other miseries, consider the state of brain disorders.

Brain birth defects routinely foredoom children to a backward life of mental retardation and custodial care. Autistic children are locked into an isolated world of compromised communication. Severe head trauma all too frequently results in paralysis, seizures, or a frightening semicomatose, vegetative state. Spinal injury can lead to quadriplegia, paraplegia, or, for the fortunate, a wheelchair. Schizophrenia, which may result from developmental abnormalities with brain neuronal deficits and faulty wiring, sentences patients to a desperate world of threatening hal-lucination, immobility, negativity, and panic. Stroke leaves the lucky sur-vivors with paralysis, blindness, inability to speak or understand, and bed-and-wheelchair existences. Midlife degenerative neurologic disease wracks postindustrial, aging societies. Alzheimer's robs people of their humanizing mentation and disrupts their emotional lives. Parkinson's converts vital individuals into rigid, limping, slow moving, depressed shadows, requiring constant care. Lou Gehrig's disease causes near total paralysis within three to five years; patients die from suffocation as their respiratory muscles fail.

Though neuropsychiatric diseases are profoundly complicated, though disorder of the brain deranges the most complicated biological system in the world, it is the unyielding, inflexible, nonreproducing, nonrenewable neuron that lies at the heart of the hopelessness of these disorders. In contrast to virtually every other cell in the body, brain neurons apparently do not divide and reproduce after development. Once damaged, the brain seems to be beyond repair. This state of brain affairs stands in stark contrast to the familiar healing and recovery that is part of everyday life. A minor cut or major wound causes skin cells to reproduce as part of normal healing that we take for granted. With only modest treatment, cells lining the duodenum of the small intestine reproduce to fill in the ulcer of the stressed person. From flu to pneumonia to anemia to urinary infection to hepatitis, healing and recovery involve reproduction and replacement of the dead and damaged cells. But brain neurons do not divide.

"Neurons Do Not Divide"

This state of affairs did not simply appear to be a misleading clinical impression based on centuries of permanent paralysis, dementia, and mutism. The sorry verdict from the bedside was confirmed by hard-edged scientific experiment performed by a neuroscientific genius of the twentieth century: Pasko Rakic. Trained as a neurosurgeon in his native Yugoslavia, Rakic had a passion to study the brain scientifically. He emigrated to the States, joined the faculty of Harvard Medical School, and rapidly emerged as a one-of-a-kind scientist.

Rakic had a gift that bordered on the miraculous. Like his peers, he examined slices of brain under the microscope to visualize cells and analyze function. Unlike his peers, though, a near-magical transformation of the two-dimensional brain slice occurred somewhere in his own brain. For Rakic, the flat slice of brain changed in space and time. He looked at a two-dimensional slide under the scope and saw the brain in its three-dimensional complexity. Living in a three-dimensional world allowed

Rakic to appreciate processes that were not apparent to his two-dimensional colleagues.

His brain also performed another minor miracle. Rakic converted the dead world of preserved brain cells on view under the microscope into a dynamic cosmos where cells moved, changed shape, and communicated. He looked at a still photograph and saw a movie—a little like reconstructing *Gone with the Wind* from a single frame. With these gifts, Rakic reconstructed the dynamics of cortical development in primates and humans. From static, two-dimensional slides of the brain, he recreated cell birth—cell migrations to distant sites and the cell interactions that result in the organ of cognition and consciousness. Innumerable experiments by astounded colleagues confirmed his vision.

Rakic turned his stereoscopic vision to neuron reproduction in the adult primate brain. He published his results in 1985 in a now-classic scientific article that carried the gloomy title "Limits of Neurogenesis in Primates."[1] The first sentence began objectively enough, "The frequently stated assumption that the adult human brain lacks the capacity to generate new neurons has never been tested in the adult of any primate species." He proceeded to test that assumption in twelve rhesus monkeys, ages six months to twelve years. To detect neuron division, he injected the monkeys with radioactive thymidine. Upon cell division, thymidine is incorporated into DNA. The "labeled" radioactive DNA is then detected in the nucleus of cells that have divided. He examined all major structures and subdivisions of the brain; for each animal, he studied approximately 100,000 cells.

Never one to mince words, he stated his observations with ringing clarity: "Not a single heavily labeled . . . neuron was observed in the brain of any adult animal." That is, neurons do not divide in the brain of adult primates. Putting an optimistic face on the whole affair, he concluded that "the brain of primates . . . may be uniquely specialized in lacking the capacity for neuronal production once it reaches the adult stage. One can speculate that a prolonged period of interaction with the environment, as pronounced as it is in all primates, especially humans, requires a stable set of neurons to retain acquired experiences in the pattern of their synaptic connectivity." Rakic was speculating that the memory required by a long, complex human life demanded stable networks of neurons, a demand that precluded the birth of new neurons and the formation of new, inexperienced circuits.

Coming from Rakic, the final word appeared to have been spoken. Case closed, door shut, QED? Almost. In a single sentence buried in the article, however, this supremely careful scientist made a prophetic statement that was ignored at the time: "Minimal radioactivity . . . was occasionally observed over a neuron." If this was a crack in the door, it certainly admitted no light in 1985. Now demonstrated by cold scientific inquiry, brain neurons do not divide in adult primates.

Bird Brains

Rakic's landmark study of primates was partly motivated by a bizarre result obtained in the distant world of birdsong. Fernando Nottebohm at Rockefeller University had been studying canary birdsong for many years. A man absorbed by his research, he was often seen running down the corridors breathlessly, late for an appointment, with a birdcage in each hand. His fellow scientists patiently waited while he meticulously set down the canaries before beginning meetings. Nottebohm was examining the acquisition of birdsong by canaries as a model to understand how the brain learns new behaviors. He was about to enter the foreign world of neurogenesis.

In many ways, birdsong seemed to be an ideal behavior for the study of learning. Enough was known to allow the framing of crisp, unambiguous questions; birdsong itself was well defined, quantifiable, and amenable to recording, preservation, and detailed analysis. Male canaries sing while females do not, which raises a host of questions that could shed light on learning. One month after hatching, infant males begin hodge-podge singing, which Charles Darwin had compared to the babbling and cooing of human babies. Gradually and progressively, they assume the adult songs. In two to three weeks the songs begin resembling those of adults. By eight to nine months of age, the males are virtuosos, singing like full-fledged adults. How does this transformation occur?

Canary song exhibited other patterns that might help Nottebohm understand learning. The song was seasonal. Males sing during the entire spring breeding season but stop singing completely in late summer. After the summer vacation, they begin learning a new repertoire in the fall for the upcoming breeding season that spring. The predictable, seasonal learning, unlearning, and relearning appeared to Nottebohm to be ideal for defining underlying mechanisms. Nottebohm set out to identify the parts of the brain responsible for song learning and then to smoke out the principles. His early studies identified a well-defined neural pathway that was used for singing. This simple system, Nottebohm found, is devoted to singing. This simple system controls a learned behavior.

Other features of birdsong only added to the intrigue. Singing depends on the male sex hormone, testosterone. Adult females treated with testosterone begin to sing. Castration of males, with the removal of testosterone, before they have learned to sing, prevents the development of adult song. During the spring breeding season, when males sing incessantly, they produce high levels of testosterone. In late summer, when singing nearly stops completely, testosterone levels fall precipitously. To add to the fascination, the brain song system undergoes dramatic changes in size as the hormonal changes occur. Here was an opportunity to relate learning, brain function, and hormonal change.

Nottebohm started by examining the song circuit during the seasons of a bird's life.[2] The revelations began. In the one-month-old toddler canary, when babbling begins, the song system of neurons is only one-eighth the size of that in the adult. As the canary progressively learns to sing like an adult, the neuron number grows to adult size. There was far more: system size changed seasonally in the adult male. It is huge in the spring, when canaries are singing frenetically, and when testosterone levels are high. Between late summer and fall, with little singing it has shrunk to half its spring size and is about as big as that of a three-month-old prepubescent bird.

Nottebohm became convinced that the size of a brain neural system determined how well the system learned the associated skill. As a corollary, he reasoned that learning somehow involved increasing the size of the controlling brain system. In addition to the developmental and seasonal changes in size, other observations supported his impression. For example, certain canaries stood out from the crowd as singing virtuosos. The operatic stars had large song systems, while less talented canaries had small systems. The real canary Carusos sported systems that were three times the size of their less gifted brethren. In females that begin singing after treatment with testosterone, the system grows in size twofold. All in all, brain system size seemed to regulate learning.

Nottebohm then attacked the big question. What was changing in the system that made it grow? What actually increased in size? Nottebohm and his student Steve Goldman, a future neurologist who was studying for a combined medical and doctoral degree, had to devise an experimental approach. Rather than comparing fall canaries with spring singers, or babies with adults, they took advantage of the hormone connection. By treating females with testosterone, they could expand the size of the song system at will, without being locked into a seasonal or developmental schedule. Normal physiology provided them with a convenient experimental paradigm.[2]

They treated a group of canaries with testosterone. They used radioactive thymidine, as did Rakic, to label any dividing cells. If testosterone enlarged song system size by causing cell division, they should detect a marked rise in labeled, radioactive cell numbers. When they examined the brains, they were dumbfounded. They experienced twin shocks. First, they saw labels in cells that looked exactly like neurons. But neurons were not supposed to divide in adults.

There was a second shock: the same percentage of neurons were labeled in the control and testosterone groups. The numbers were equally astounding. One-and-one-half percent of the neurons were labeled each day. At that rate, the number of neurons would double every forty-nine days! The neuronal division, if real, occurred in normal control canaries as well as ones that were testosterone-treated. Nottebohm and Goldman were left with the inescapable conclusion that neuronal division is a normal event in the adult canary brain.[2]

It would be the most charitable of understatements to say that the results were greeted with skepticism by the scientific community. As Nottebohm traveled from conference to conference presenting his results, he was deluged by a hail of criticism. Distilled to its essence, the central question was "How do you know the labeled cells are really neurons?" Many cells can look like neurons under the microscope. Nottebohm returned to his laboratory and examined the new cells under the powerful electron microscope. He and his colleagues found synapses forming on the newly born, supposed neurons. But skeptics noted that synapses form on many cells that are not neurons.

It became apparent that the only way to convince the skeptics, and himself, that the new cells were neurons was to define their electrical properties. Neurons exhibit characteristic electrophysiological traits not seen in other cells. This study, however, represented a monumental undertaking. Using microscopic electrodes, newly divided cells had to be located in the song system. The electrodes had to be inserted into the cells in the intact brain, and electrical activity had to be recorded under varying conditions. This required the highly specialized electrophysiological expertise of John Paton at Rockefeller. After the electrical tests, Nottebohm and Paton had to verify that the cells from which they were recording were indeed newly born. Each cell that was characterized electrophysiologically had to be analyzed for thymidine labeling, an awesome yet rewarding task. The newly dividing cells in the adult brain were, without doubt, neurons.[2]

Now that Herculean efforts established that new neurons are born in the adult canary brain, many other pressing questions vied for attention. One overarching question concerned the generality of neurogenesis in adults. One question that is subordinate to this general one is whether the song system is a freak of nature, harboring the only group of newly born neurons in the brain. Nottebohm looked elsewhere and found new neurons in multiple areas of the cerebral hemispheres, in the cerebellum, which controls balance, and in the brainstem, which connects the hemispheres and the spinal cord. Since tens of thousands of new neurons are produced throughout the brain every day, neuron production is widespread in the canary's brain.

Another related question is whether there is something anomalous about the canary itself. Is the brain of this songster an exception specializing in the production of neurons? No, again. Nottebohm detected new neurons in the brains of parakeets (budgerigars) and doves. In sum, new neuron production occurs in many areas of the adult brain of many species. The extraordinary phenomenon of neurogenesis in the adult brain is generalized in birds.

New neuron production is widespread, then. Tens of thousands of new neurons are produced daily in the adult bird brain. Nottebohm estimated that, at this rate, the number of neurons in the canary brain would double every forty-nine days. Yet, the bird brain does not undergo ever-expanding growth throughout life. The song systems of Nottebohm's one-year-old female canaries (treated with testosterone) contained about 15,000 neurons, the same as the two-year-olds. Their brains did not grow like a cancer, ultimately dwarfing the rest of the bird in an encephalized, Martian-like, caricature. It was apparent to Nottebohm that ongoing neuron death must accompany ongoing birth. The singing male in the spring, for example, has about 40,000 neurons in the system, but only 25,000 a few months later in the fall. Massive neuron death occurs all the time. Neurons are constantly turning over in the adult brain—neuron numbers represent a balance between neuron production and neuron death. Growth of an area such as the song system is caused by a relative increase in birth versus death. Shrinkage after the season of song is due to a preponderance of death. The phenomenon of neuron turnover and the contributions of birth and death at any time account for the constancy of brain size and the growth of specific areas.

Yet another question is, where are the labeled neurons actually born? Nottebohm and Goldman had identified labeled neurons that had already divided; they had not caught the progenitor cells in the act of dividing. Yet this question is crucial to understanding which parts of the adult brain have the power to regenerate. Dreams of distant strategies of brain repair and recovery of function after stroke or trauma depend on locating the newborn nursery. Nottebohm and Goldman injected labeled thymidine and examined the brains after brief intervals to catch cells in the act of dividing. They succeeded. The neurons were born in a narrow zone lining the lateral ventricles, the normal cavities deep in the cerebral hemispheres. This "subventricular zone" is also the region where neurons are born during the normal embryonic developmental period.

While neurons are born in the subventricular zone, they end up in the song system, among other places. The neurons migrate vast distances to reach their destination in adult brains, just as they do in embryonic brains. Cell division and migration are two key processes that underlie the brain's development. In a sense, then, Nottebohm and Goldman discovered that developmental events, specifically cell division and migration, persist in the adult brain. If these processes could be controlled in the adult, regeneration and recovery after illness or injury are conceivable. The stakes could scarcely have been higher.

Mammalian Neurogenesis Revisited

As 1990 approached, the brain birthday dilemma had become embarrassing. Neurons were continually born in one group of adult vertebrates, birds, but not in others—primates. Yet the headline news in all the rest of biology was the unity of life: From the Human Genome Project to cell biology to developmental biology, life is conserved by evolution. From roundworms to fruit flies to sea snails to humans, genes are shared, developmental processes are shared, neural mechanisms are shared. Even intelligence traits are shared. It was implausible that a process as basic as cell division occurs in one closely related group but not in others.

There had been scattered studies over the decades that suggested that the birth of selected neurons does occur in some adult brains. Although these instances were regarded as exceptions to the rule, they indicated that, in principle, neuron production is possible. For example, it was demonstrated that a population of neurons in the hippocampus, a center for learning and memory, are produced throughout adulthood.[3] The observations had been noted over the years in mice and rats, the traditional laboratory animal models. Nevertheless, the phenomenon of neuron production was intellectually compartmentalized by the scientific community as an exception to the rule.

In another example, we had known for decades that neural cells in the olfactory (smell) system of mammals constantly turnover in the adult. New cells are continually produced, and replace naturally dying cells. These cells were also regarded as an anomaly, exceptions to the rule that neurons are not produced in the adult.

In the absence of an appropriate conceptual framework, self-renewing neurons could only be regarded as an oddity of little general significance. This selective indifference, and even willful ignorance, conforms to the semi-facetious observation made decades ago by pioneer quantum theorists. Turning the scientific process on its head, the assertion was made that experimental "facts" could not be accepted until verified by theory. Neuroscientists had no theory into which adult brain neurogenesis could fit. Discoveries, and often entire bodies of scientific work, are ignored for years, until the world catches up. As 1990 approached, neuroscience began catching up.

Several intersecting currents moved the problem of neuronal cell division to the center of scientific awareness. The new appreciation of the role of growth factors in the economy of life was pivotal. Growth factors prompted division of a wide variety of other cell types; they also exerted multiple effects on neurons. Since neurons were now known to change functional states in response to these external signals, perhaps neuronal mitosis was conceivable. In a general sense, neurons were being seen as "plastic," flexible, ever-changing microcosms—and perhaps they could divide. At last the art of neuronal cell culture had reached the point where questions about division could be answered in a Petri dish. The intellectual stage was set, and the lab techniques were available to initiate intensive experimentation.

Scientists in the United States, Canada, Germany, and Australia were simultaneously hunting for dividing neurons. This activity couldn't be termed a race because experiments were being performed independently, without awareness or communication among the investigators. No, this was simply a scientific search whose time had come.

Neuronal Division in Culture

One significant avenue of inquiry employed the powerful cell culture approach to analyze development. In several laboratories, the processes governing growth of neurons were being characterized without focus on cell division itself. The possibility of actually studying neuronal division emerged unexpectedly. In the late 1980s, Emanuel (Manny) DiCicco-Bloom was examining the growth of sympathetic neurons. These neurons were popular models of development since they were easily accessible as part of the peripheral nervous system, lying outside the brain. They were extremely important physiologically because they were pivotal in the classic fight-or-flight response. Also, because they depended on NGF for survival and development, they were long used to study NGF actions. DiCicco-Bloom ultimately took the sympathetic neurons in a new direction.

DiCicco-Bloom himself had already moved in a few new directions. As an undergraduate, he transferred from Georgetown University to Princeton, where he pursued several interests simultaneously. He studied pottery with a recognized master at Princeton and became a gifted potter. His mentor was the revolutionary potter and sculptor Toshiko Takaezu. In the post–World War II period she transformed pottery from traditional devotion to utilitarianism to abstract sculpture. The diminutive Takaezu created towering ceramic spheres and ellipses fused with colorist glazes. The shimmering surfaces often suggested calligraphic forms, perfectly accenting the piece. Much of her work synthesizes the disciplined freedom of Zen with the origins of pottery in the deep Neolithic period. Her organic forms influenced DiCicco-Bloom, who created biomorphic pots and cups with lifelike glazes. DiCicco-Bloom's pottery gave expression to his love of biology.

His pottery sculpture complemented his longstanding interest in fine art, and he worked assiduously on his drawings and draftsmanship. At the same time he managed to nurture his passion for biology. He undertook a student research project at Princeton: the development of the nervous system in frogs. DiCicco-Bloom was interested in how nerve fibers find their proper targets during development, a classic problem. He transplanted patches of skin to distant locations in the developing frog and followed nerves to determine whether they grew to the correct spatial location but incorrect patch of skin, or whether they grew to the correct, transplanted patch of skin in an improper location. While time for student projects was short, and final answers were not available, the work, along with his other special gifts, placed a novel stamp on his career. His scientific work was endowed with a rare esthetic. His future publications were marked by magnificent, illustrative photographs and drawings of neurons, and his lectures were always accompanied by imaginatively conceived slides. DiCicco-Bloom's approach anticipated the computer graphics revolution.

As his Princeton undergraduate tenure drew to a close, he had to make a decision. The worlds of pottery, sculpture, fine arts, science, and medicine beckoned. He elected medicine and gained acceptance to Cornell Medical College, but maintained an interest in his other loves. In the summer after his first year in medical school, he joined the Laboratory of Developmental Neurology and rapidly distinguished himself as a scientist with unique intellectual and esthetic gifts. His experiments on sympathetic neurons in culture laid the foundation for pioneering work on neurogenesis.

His single summer project was so inordinately successful that he was compelled to take a year off from medical school to pursue his laboratory experiments in the context of his medical studies and artistic perspective. Owing to an unaccountable act of the muses, the laboratory was located crosstown in Manhattan from the famed Art Students League that had guided generations of artists in New York. DiCiccio-Bloom enjoyed the best of both worlds, although he had a bit of a scheduling problem. He synchronized his experiments with art classes, rushing crosstown, either east or west, to complete a sketch of a model or to initiate an experiment. The combination worked: he drew magnificent sketches and contributed mightily to a new field of neuroscience.

DiCicco-Bloom began studying young sympathetic neurons in culture. He knew from previous experiments that sympathetic neurons migrate to their final locations in the embryo between 11 and 13 days of gestation. DiCicco-Bloom, interested in early development, microsurgically removed neurons from the thumbnail-sized, 14-day rat embryo—a technological tour de force. While growing neurons in culture, he studied the intimate details of growth much earlier than had ever been achieved. He hit pay dirt: the young neurons provided entirely new views of neuronal development.

One problem with sympathetic neurons, models of development that they are, is their dependence on NGF for survival. While the responses of sympathetics to NGF led to many key insights to the growth factor's actions, it also was potentially confusing. Were observed effects simply due to high doses of NGF required in culture, or did they truly reflect normal development? Stated more simply, were scientists studying the effects of a naturally occurring drug, NGF, or were they getting to the heart of normal development? Manny's system of embryonic neurons provided a route out of the dilemma: young sympathetic neurons did not require NGF. DiCicco-Bloom discovered that sympathetic neurons become dependent on NGF later in development. This opened the way to understanding how neurons develop in the absence of NGF. Neurons developing in the absence of external growth factors enabled DiCicco-Bloom to define intrinsic mechanisms that regulate sympathetic development. The new culture system also allowed DiCicco-Bloom to grow younger and younger neurons, permitting new approaches to the problem of cell division, an early phenomenon.

Previous culture work had used older neurons and cell division did not occur. It had been assumed that, unlike other cells, neurons exhibited their usual ornery characters and did not divide in culture. This new system of extremely immature neurons appeared to be tailor-made to revisit the question of neuronal division (mitosis). DiCicco-Bloom used even younger neurons that he had placed in culture 13.5 days after conception.

To maximize the chances of catching a neuron in the act of dividing, combinations of growth factors, known to induce division in more reasonable cells, were added to culture dishes.[4] Radioactive thymidine, incorporated into the DNA of dividing cells, was used to detect cell division. In dramatic contrast to all previous experiments with mammalian neurons, many culture dishes of these young neurons contained cells with the telltale thymidine in their nuclei. Their identity as neurons was rapidly confirmed. By calculating how many labeled neurons were in each dish, investigators readily saw that a specific family of growth factors markedly enhanced neuronal division. It was the insulin family of growth factors that maximized division of these young neurons under such conditions.

This breakthrough provided critical insights. Most important was that normal, noncancerous, mammalian neurons can divide in culture. No internal instructions forbid these neurons from reproducing in cell culture. In turn, cultures can elucidate the regulation of neuron division, a previous impossibility. But the work provided another clue. Discoveries with the insulin growth factors indicated that external signals could elicit neuronal division. Consequently, neuronal division was not enslaved to an immutable, inaccessible internal clock. If environmental neurohormones can make neurons divide, external treatments might someday do the same.

Though the world of the nervous system now seemed a little less hopeless, several essential questions pressed for answers. Are brain neurons also capable of dividing in culture, or is this capacity restricted to peripheral sympathetic neurons? Is neurogenesis restricted to early development in mammals?

By using neurons from the developing cerebellum, the brain region governing balance, DiCicco-Bloom found that brain neurons, too, divide in culture.[5] In sum, developing brain and peripheral mammalian neurons divide in the cell culture environment. The power of culture could now be focused on elucidating the molecular controls of neuronal division. That knowledge might someday foster regeneration and recovery after brain illness and injury—the Holy Grail.

Adult Brain Revisited

The all-important question of neurogenesis in the adult mammalian brain was being investigated in other laboratories. Twenty thousand miles away, in Victoria, Australia, Perry Bartlett and his colleagues cautiously reapproached the adult mammalian brain and used the mouse as a model.[6] They noted that olfactory (smell) neurons are continually produced in adult brains of many species. They further noted that granule neurons of the hippocampus, a memory center, are also born after birth. Finally, they reviewed Nottebohm's writings on neuron production in adult birds. The role of testosterone was particularly salient for Bartlett and his colleagues. They had found that growth factors stimulated proliferation of neurons from *developing* mouse cells, analogous to the actions of testosterone in the adult bird brain. Emboldened by this line of reasoning, they asked whether the adult mouse brain also contains cells that can respond to growth factors by dividing and producing neurons.

They used the time-honored Rube Goldberg approach of combining odd bits and pieces of previous work to fashion their own unique experiment. They began with adult mouse brains. They used the cortex, the thinking conscious areas, the hippocampus, a memory center, and the base of the brain, which is devastated by Alzheimer's. The cells were placed in culture to grow. Drawing on their own experience with growth factors in development, they added fibroblast growth factor to enhance the possibility of inducing cell division. In some cultures they employed other growth factors to maximize the chances for success. And they added a new twist. Some cultures were exposed to astrocytes, star-shaped cells that support neurons in the brain. These cells may make other factors important to the health and well-being of neurons.

The recipe worked. Cells in the cultures divided, and produced new neurons.[6] Optimal production of neurons required the presence of the fibroblast growth factor and the astrocyte factors.

Was neuronal production restricted to specific areas? To examine this question, they subdivided isolated areas prior to culture and found that multiple regions contained cells that could divide and give rise to neurons. The conscious, thinking cerebral cortex, the remembering hippocampus, and the Alzheimer's-susceptible base of the brain all gave rise to new neurons. This widespread ability reproduced the discoveries of Nottebohm in the canary brain, indicating that neuronal production transpires in many areas simultaneously.

In principle, the discovery in adults paralleled the observations of DiCicco-Bloom in developing neurons: Growth factors induced division of cells that result in the production of neurons. The all-important difference was that Bartlett's experiment indicated that the *adult* mammalian brain contained cells that retained the ability to divide and produce neurons. As the authors observed, "The most exciting ramification of these findings is the possibility of stimulating the neuronal precursors . . . to replace diseased and damaged neuronal tissue."[6] They were already dreaming of stimulating the resting progenitor cells to divide as replacement treatments for neurons dying from illness or injury.

The Australian group excitedly submitted their scientific manuscript for critical review and publication. While the work was being reviewed, however, *Science* published a paper that also focused on neurogenesis in the adult brain. The work was performed halfway around the world, at the University of Calgary in Canada. Neuronal production in the adult brain was clearly coming of age.

Neurospheres

Brent Reynolds and Sam Weiss in Calgary used a related approach to search for neuronal production in the adult mouse brain, though their driving motivation was different.[7] They were particularly interested in a different member of the vast alphabet soup of growth factors, epidermal growth factor (EGF). Weiss and Reynolds were struck by several observations. EGF is a potent inducer of cell division in a wide variety of non-neuronal cells. EGF was known to stimulate wound healing and the regeneration of a variety of adult tissues, including skin, liver, and intestine. EGF also caused cell division in the developing brain. Finally, EGF itself and its receptor had been detected in adult rodent and human brains. They began experiments to determine whether there were cells in the adult brain that could divide in response to EGF and produce neurons.

They began by isolating cells from the striatum in the adult mouse brain. This area controls automatic motor behaviors, such as bike riding (in humans, not mice), and is affected in Parkinson's disease. They placed the striatal cells in culture with EGF. Though most cells died after one week in culture, spheres of proliferating cells remained in the dishes. Under particular growth factor conditions, the dividing cells gave rise to neurons.

Reynolds and Weiss turned their attention to the dividing cells. What exactly was their identity? The dividing cells contain a protein—a hallmark of a critical type of embryonic cell. This special type of cell, termed a "stem cell," can give rise to neurons and to non-neuronal cells and is self-renewing. Stem cells have the unique ability to generate many cell types and are therefore keys to brain development. Persistence of stem cells in the adult brain was unexpected, raising remarkable possibilities for regeneration.[8]

True stem cells have the potential to give rise to every cell type in the brain. They are therefore termed "multipotent." They are the true Eves (or Adams) of the developing brain. In theory, stem cells can divide and generate the hundreds of types of neurons, and the multiple types of non-neuronal cells that comprise the brain. Moreover, since stem cells are self-renewing through division, they are an inexhaustible source of all cell types. Stem cells could be cultured, allowed to multiply as stem cells, and then frozen and stored for future use. By growing them with appropriate growth, survival, and differentiation factors, brain neurons of any type could theoretically be generated. Though subject to qualifications, the presence of dormant stem cells in the adult brain implied that a brain could be rebuilt in its entirety after illness or injury.

Several irresistible, experimental first steps have already been taken in the live animal. One question deriving from the discovery of stem cells concerns the effects of treatment of live animals with EGF. If EGF increases stem cells and cell division in culture, can it increase cells in the living brain? EGF was infused in the brains of living mice for a week and stem cells were noted. The growth factor caused nearly a twentyfold jump in the number of cells. So we are obviously on the threshold of a new era. We can entertain the fantasy of rebuilding a brain using growth factors and stem cells. Though just a dream now, we have the knowledge and tools to articulate the problem.

Primate Brain Revisited

The gathering scientific momentum encouraged a reexamination of the question of neurogenesis in adult primate brain. Bruce McCewen and his student, Elizabeth Gould, at Rockefeller University, examined neurogenesis in rodents; Gould subsequently extended the studies to primates, including marmosets. They were able to detect the birth of neurons in the hippocampus.[9] It had been known for a number of years that cells were born in the adult hippocampus of rodents. Gould moved to Princeton to assume a faculty position, and continued her studies by extending them to higher primates. She rapidly examined old world monkeys, our closest relatives. Neurons were born in the hippocampi of adult, old world primates. This discovery, however, appeared to contradict the classical work of Rakic, discussed above. Why was there a contradiction?

Rakic and his postdoctoral trainee, David Kornack, examined old world monkeys, but could find no evidence of neurogenesis. This was potentially the gathering of a scientific controversy with very high stakes. A major conflagration was avoided, and agreement was reached, when Rakic and Kornack used a more advanced technique, and were also able to detect neuronal birth in the adult old world monkey brain.

The Human Brain

The rush of scientific discovery was raising the heretical possibility that brain neurons were a renewable resource after all. If it was true of mice, was it true of men? Steve Goldman, Nottebohm's former student, now a neurologist at New York Hospital Cornell Medical Center, joined forces with David Pincus, a neurosurgical trainee at Columbia Presbyterian Medical Center, to approach the human brain.

Pincus had been a student of DiCicco-Bloom and played a pivotal role in showing that growth factors govern neuron division in culture. As a medical and doctoral student at Cornell, Pincus had discovered that environmental signals caused neurons to divide, that electrical signals and growth factors induced division. In principle, then, Pincus's research established that nurture interacted intimately with the neuron's nature to regulate mitosis, a formidable achievement for a distinguished scientist, much less a medical student. Pincus was no ordinary student. Tall, handsome, articulate, and poised, with perspective and judgment befitting a senior faculty member, he was more colleague than neophyte. One notable Sunday his characteristic grace failed as he moved stiff-legged through the laboratory, clearly in physical distress. Only on close questioning did he allow that he had just run the New York City Marathon of twenty-seven miles—and had raced back to the lab to complete his experiment. It would have taken far more than a marathon to impede the progress of Pincus.

After receiving his doctorate, he returned to medical school to complete his training. He then embarked on another marathon, a seven-year program to become a neurosurgeon. As though this activity was not arduous enough, David decided to subspecialize in pediatric neurosurgery. David was to become the only neurosurgeon in the world to simultaneously hold the keys to neuronal division and perhaps regeneration.

Pincus and Goldman had a straightforward plan. In the neurosurgical operating room, Pincus frequently was compelled to remove small sections of a patient's brain. The patients suffered from intractable epilepsy that caused devastating seizures even on high doses of medication. To alleviate the symptoms, the brain area that generated the seizures had to be excised. The area removed included part of the cortex. Their plan was to place the human cortical cells in culture and try to find neuronal division.

It is a complex journey from operating table to culture dish. The approach called for combining delicate surgery on the brain with the fastidious demands of brain cell culture. The brain cells removed at surgery had to be treated rapidly to maintain viability. They had to be immediately transported to the laboratory under sterile conditions. As soon as they arrived in the laboratory, the cells had to be prepared for culture. Time was of the essence. The surgical team in the operating room had to be precisely synchronized with the culture team in the laboratory. The careful planning paid off. Cells removed from the patients' brains grew in culture. The ambitious project of searching for neuronal division in the human brain could now begin.

Although the brain tissue removed at surgery yielded far fewer living cells in culture than that from animals, enough survived to permit experimentation. Pincus and Goldman subjected every culture dish to exhaustive microscopic scrutiny in search of a neuron labeled with radioactive thymidine, the sign of mitosis. It was necessary to search through many cultures of human cells to identify any neurons at all. The overwhelming number of cells in culture were non-neuronal support cells. These were distractions, of little interest at the moment. Most of the neurons they did detect exhibited no radioactivity in their nuclear DNA. Finally, extensive examination revealed isolated neurons with labeled DNA. Neurons from the human brain were reproducing in culture. What was true for birds, mice, and rats was true for the human brain.[10] The distant possibility of renewing neurons in the human brain was one step closer.

Several serious qualifications came into play. Pincus and Goldman were working with epileptic, diseased brains. Were they simply detecting an abnormality associated with disease? Their patients were heavily medicated with anti-seizure drugs. Were the drugs inducing neurons to divide? The brain tissue was removed from patients under anesthesia at the time of surgery. Was there something about the anesthesia that induced neuronal division? They could examine only a few patients initially. Were the findings applicable to all human brains? Caveats aside, the groundwork was now laid to potentially treat brain disease in the future.

Astounding Prospects

The general significance of the discovery that brain neurons can reproduce in the adult mammal, including humans, may seem all too obvious. But the revolutionary implications of Pincus and Goldman's work are highlighted by findings in related fields. One particularly hot area, termed "brain transplantation" by the media, or grafting of cells to the brain, bordered on the miraculous.

The scene shifts back to Sweden. Indeed, in neuroscience, the scene never shifts too far from Sweden. This tiny country of some 7 million souls has been in the vanguard of neuroscience for most of the twentieth century. If one were to calculate neuroscientific discoveries per capita (which may be as valid an index of national wealth as GNP, GDP, trade surplus, wars won, world football cups captured, or movies exported), Sweden would still be in the age of Viking ships, dominating the world. Let's look at the bucolic, ancient city of Lund in southeastern Sweden.

The University of Lund is a medieval institution with tree-lined paths and enough bicycle traffic to bring the strolling, contemplative scientist precipitously back to earth with a big bang. The university is the home of Anders Bjorklund, neuroanatomist, developmental neuroscientist, histologist, neural cell biologist, and innovator of the first rank. Short and spare of build, with dark hair, piercing eyes, and a crisp articulateness, Bjorklund always appears to be making new connections among disparate fields. He synthesized his variegated interests to help create the new endeavor of transplantation of cells to the brain. By transplanting cells to brains of different ages, we might understand how the brain environment changes during development and alters cell fate and function. Numerous parallel questions in the developmental biology of the brain can be approached with this strategy. By transplanting cells of different types to brains at the same stage, it might be possible to assess the intrinsic potentials of different cells in similar brain environments. Bjorklund played a central role in launching this approach. With his collaborator, Fred (Rusty) Gage, he used grafting to foreshadow entirely new approaches to treating brain disease.

Gage worked as a collaborator in Bjorklund's laboratory in the mid-1980s. Gage was already a true international. He was an American who grew up in Rome, where his father was one of the first Merrill-Lynch international account executives. Gage received his doctoral degree from Johns Hopkins University in Baltimore, studying behavior of rats with the famed animal psychologist David Olton. He moved to Lund to investigate the brain basis of behavior by using Bjorklund's innovative approaches.

Gage's family had already made unorthodox contributions to neuroscience. His great grandfather, Phineas Gage, worked on the construction of the transcontinental railroad in nineteenth century America. As an engineer-worker, he used pipes filled with explosives to clear the way for ties and track. Phineas Gage was known as a man of exceptional politeness and probity. He was a model of the hard working, clean living, nineteenth century man. One fateful day he tamped down a charge-filled pipe and it exploded unexpectedly, sending an iron bar through the frontal lobes of his brain. He was carried from the scene by coworkers who held little hope. He lived, but he was a changed man. The proper Phineas Gage became a dissolute, irresponsible, hard drinking drifter, who died a few years after the accident. His brain was preserved and continues to be analyzed to this day. In this curious way, Phineas Gage made major contributions to our understanding of the role of the frontal lobes in the genesis of personality.

Gage, working with Bjorklund, made contributions of another sort. Interested in the mental deficits that occur with aging, they focused on rats who had reached the ripe old age of two to two and one-half years.[11] Employing a battery of behavioral tests, they identified a subgroup of elders whose performance indicated near-senility. They focused on the basal forebrain—which degenerates in Alzheimer's disease—and its connection to the hippocampus, a memory center. They tested the rats' spatial memory, which is controlled by this basal forebrain-hippocampal system. The aged subgroup could not learn to negotiate a maze for rewards. The deficit in spatial memory resembled that of Alzheimer's (as suffered by Enoch Wallace). Using published work on Alzheimer's brains, they reasoned that one key deficit might lie in the basal forebrain neurons; so they transplanted basal forebrain neurons to the brains of the memory-impaired rats. After allowing time for recovery from surgery, they trained the rats on mazes and then tested them to assess learning and memory of spatial tasks. Transplantation of new neurons dramatically improved learning and memory! In one breathtaking leap, a new way of thinking about treatment of mental deficits appeared on the horizon.

Against this background, the discovery that new neurons can be produced in the adult brain could hardly be of greater significance. Cell transplantation therapy requires cells. Theoretically, stem cells in the patient's own brain could even constitute the source. At the least, cells could be grown in culture, expanded in number and then transplanted. The birth of new neurons could hold the key to the feasibility of cell therapy in the brain.

Adult brain neurogenesis may affect other diseases as well. Alzheimer's is not the only brain disease of relevance, and Lund is not the only center of excellence in Sweden. Several miles to the south, in Stockholm, Lars Olson was performing his own brand of scientific magic at the Karolinska Institute. Among many interests, Olson had one eye fixed on Parkinson's disease.

A small group of medical school classmates at the Karolinska had revolutionized neuroscience, carrying out the scientific program initiated by their mentor, Nils-Åke Hillarp. Hillarp, with Bengt Falck, had invented the fluorescence technique that revealed the neurotransmitter signals norepinephrine, epinephrine, and dopamine (collectively known as "catecholamines") to be visualized under the microscope. The technique was used to map transmitter signal systems throughout the nervous system.

But tragedy struck, and a young Hillarp lay dying of cancer. He summoned his students to his bedside and one-by-one he assigned each student a program to ensure that the campaign did not falter in his absence. Olson, a colleague of Hillarp's minions and one of the indirect beneficiaries of the legacy, became an expert in the microscopic study of catecholamine neural systems. One catecholamine neuron group in the brain was of particular relevance since its degeneration resulted in Parkinson's disease.

A rat model of human Parkinson's disease was developed by selectively destroying critical neurons (which, for convenience, we will call the "Parkinson's neurons"). The rats exhibited the slowness, awkward gait, muscle rigidity, and tremors of the human disease. Olson's long-standing interest in neuronal development, catecholamine neurons, and neuronal function in different environments served him well as he attacked the problem. He transplanted catecholamine cells to the brains of rats with experimental Parkinson's disease.[12] The rats improved dramatically: their movements became rapid once again; their gaits were close to normal; the rigidity and tremors largely melted away. Again, cell therapy had rescued the failing brain. In this instance, Olson was compelled to use substitute cells for the true Parkinson's neurons because the latter were mostly inaccessible in the brain and available only in minuscule amounts. Instead, he employed cells from the adrenal gland whose signals are related to the true Parkinson's neurons.

Hence we see another instance where stem cells would be invaluable. With the proper recipe of growth and differentiation factors, stem cells can be transformed into true Parkinson's neurons. The possibilities, and requirements, for brain stem cells were dizzying. Though scientists talk about it in hushed tones, prospects for neurogenesis in the adult brain have changed the landscape. The unthinkable possibility of regenerating neurons after brain illness and injury had at last become a subject of rational discussion among sane scientists.

❖

Enoch's disordered thinking entered an ominous phase with his jealousy of Charlie Bachrach, his most trusted and inseparable friend since college. Enoch became obsessed with an imagined Sally-Charlie affair, and exploded in a rage, confusing and wounding wife and friend. The fragmentation of thinking and communication in Enoch's brain is most probably attributable to pathology of non-neuronal support cells, as well as of the neurons themselves. The support, or glial ("glue") cells, the subjects of study for a century, are now known to maintain the brain environment, provide trophic support, insulate nerve fibers allowing electrical conduction, and act as immune-inflammatory bodyguards. The glial cells also appear to critically facilitate neuronal communication, fostering the information flow, so deranged in Enoch.

❖

7

BRAIN GLUE

❖

ENOCH AND CHARLIE Bachrach had been buddies since college, living in adjacent dormitory rooms, joining the same drama club, dating the same group of women, and, incidentally, majoring in economics and political science before the requisite tour of duty in law school. Since they both topped off at an even 6 feet with slim, if not willowy, frames, they could share shirts, trousers, belts, and ties, a convenient way to expand their wardrobe to enviable proportions. Between them, they could drape themselves appropriately for most social events, whether the informal mixer or the black tie dinner dance at Christmas. Though they finally went to different law schools after much deliberation, they maintained weekly phone contact, spent vacations canoeing together, and went on grand tours of Europe, riding the rails from Paris to Florence to Rome. They were best men at each other's weddings, godfathers to the resulting progeny, and participated in each other's family outings. Each summer, two to three weeks were devoted to a joint family odyssey, often a camping trip, replete with cranky children and distressed pets. Though they were at different firms, the close world of Wall Street solidified their commonality of interests. Friends joked that Enoch and Charlie were so close they no longer needed to engage in the tedium of verbal communication.

Sally was mildly surprised when Enoch asked one morning about the day that she and Charlie had planned to visit the Armory for the annual antique show. Antiques were an addiction she shared with Charlie; they had always bored Enoch.

"You've been spending a lot of time with Charlie."

"What do you mean?"

"The whole Armory could be covered in two hours. What are you two doing the rest of the day?"

"We'll have lunch, I suppose, and maybe visit the museum."

"That's a lot of time."

"Enoch, what in the world are you talking about?"

"You're spending a lot of time with Charlie."

"Enoch, what *are* you talking about?"

Sally attributed this strange little interchange to Enoch's recent distractibility. But the subject did not stop with this conversation.

"Sally, he called again."

"Who?"

"Charlie."

"Well, what did he have to say?"

"He was calling for you, Sally. I thought you'd know."

"Enoch, how would I know?"

"He's been calling for you a lot. What are you two talking about all the time?"

"I haven't spoken to him for weeks, since we made plans for the antique show."

"Then why is he always calling for you?"

"Enoch, what on earth are you talking about?"

"You speak to Charlie more than you speak to me."

"Enoch, this is crazy. What are you saying?"

"You're spending a lot of time with Charlie. That's all."

Enoch's preoccupation with Charlie and Sally continued to escalate, and reached a bewildering height with the sundress affair.

"What are you wearing to the antique show?"

"I thought I'd wear the white and yellow sundress."

"You haven't worn that with me for a long time."

"It's perfect for this sticky weather."

"That's a pretty revealing outfit. Everyone will be staring at you."

"Enoch, it's an ordinary sundress."

Without warning, Enoch launched into an explosive crescendo. "You talk to Charlie all the time. He calls you every day. All you think about is Charlie. Now you're going off with him in a sexy dress, in public. Every time I mention Charlie, you give me the silent treatment. I don't even know what you two will be doing all day. What's going on between you and Charlie? How do I know where you'll really be? You could be all over the city. And now you're going to wear the sundress! Who dresses like that for an antique show? What's going on? For all I know, you and Charlie are running off somewhere."

"Enoch!"

The diatribe ended in a shout, breaching a formerly inviolate boundary. The damage was done. The explosion took them both by surprise. They were confused. Neither understood what had just happened. But this was only the beginning, only a hint of the rage to come. Enoch and Sally had entered a new, lacerating phase.

Many technical and lay terms are used. None captures the personal pain. Delusional thinking, delusional systems, mild paranoia, catastrophic reactions. Perhaps Dr. Alois Alzheimer captured it best a century ago in Germany when he described the jealousy of the wife, his first patient with the disease. Her cardinal symptom was jealousy, not memory loss, not learning deficits, not speech difficulties. Jealousy was the hallmark of her disordered thinking.

❖

While neurons form the electrically conducting, communicating brain circuits, they actually comprise a minority population. There are ten times as many non-neuronal, support cells, termed "glia," or glue cells (from the Greek) in the brain. For the longest time, their function has been a mystery, though their contribution to disorder has been apparent. The deleterious effects of glial scars after injury has long been recognized.

Jeremy Soreno, a twenty-two-year-old sophomore at the university, took the familiar, winding River Road back to the dorm at night after finishing his waiting shift at the restaurant. February second was a particularly bad night: it was snowing just enough to obscure vision, the blacktop was icy, and the bald patches on his tires frequently skidded. The careful, aware Jeremy did not ask too much of his old blue Volvo sports car, easing into turns and maintaining a steady 30 miles per hour on the rare straightaways. Even so, it was impossible to avoid skids entirely. The road was deserted at this hour, and there was no danger of encountering blinding, oncoming headlights. The last scene Jeremy remembers was skidding, out of control, toward the massive trunk of the dead elm at the corner of the Grigg's farm.

Jeremy awoke from his brief minutes of stupor in an intensive care unit of the community hospital. His head and neck ached. Then the drive home, the snowstorm, and the elm tree flooded into consciousness, and he pieced together his situation. Though he felt drowsy and confused about his location, he had the presence of mind to run a quick spot-check. He knew his name. Though hampered by hospital restraints, he was definitely able to move all four limbs, and could wiggle all his fingers and toes. As best he could tell, he had feeling throughout his body. He was able to see in the dim room, and could hear the beeps of the EKG monitor distinctly. With that much intact, he could relax, look around, and reconnoiter.

Within seventy-two hours Jeremy was alert and oriented, his neurologic exam was normal, and the laboratory tests, including skull x-rays and MRI, were negative. He walked out of the hospital four days after the accident, carrying a diagnosis of concussion and closed head injury, but actually felt normal in all respects. He returned to classes full time, without event.

Jeremy experienced his first seizure six weeks after the accident. As he began arising from the wooden cafeteria chair, he suddenly fell to the floor. His body thrashed violently and his arms and legs flexed and extended in a hyperkinetic, hideous, chaotic fashion. Pink foam dribbled from his mouth and he was incontinent of urine and stool. His friends were transfixed, horrified. The entire episode didn't last longer than thirty seconds. The thrashing stopped, he gradually regained a confused, garbled consciousness in minutes, and he was lucid in an hour. Over the next two weeks, he experienced one or two seizures each day, before treatment was definitively instituted.

Jeremy was suffering from delayed, posttraumatic epilepsy. A glial scar had been forming in his cerebral cortex. By six weeks it had begun exciting adjacent neurons, resulting in abnormal firing and seizures through unknown mechanisms.

What do these glia normally do? Certainly, they were not placed in the brain to cause seizures. What is the role of glia in normal brain function?

❖

So far my analysis of brain function has concentrated on neurons. These cells, after all, are the brain's signaling elements. The neuroscientific community long ago established the primacy of communication in brain function. Neurons generate electrical impulses that convey messages to other neurons. Since the brain processes information, neurons are undoubtedly appropriate foci of attention. When you view a van Gogh, impulses from your retina are conveyed to the visual cortex at the back of your brain, and you perceive the image. When you smell a rose, the olfactory nerves transmit impulses through several way stations from nose to olfactory cortex in the temporal lobes on the side of your brain; these nerves elicit pleasurable sensations and associations. When you walk and chew gum, the motor cortex fires impulses down the spinal cord and out to the muscles, which contract and relax. The focus on neurons has been abundantly rewarded with major insights to brain function. Before I close accounts with brain reality, though, I want to touch on a few interesting facts.

The Glue Cells

Neurons are actually a minority population of brain cells. There are approximately ten times as many non-neuronal support cells as there are neurons, so if you have roughly 1 trillion neurons, your brain has 10 trillion support cells. Though numbers alone hardly confer status or prestige, it is difficult to ignore 10 trillion individuals, however undistinguished. Yet 10 trillion brain cells have led a backward existence, hidden in the shadows for nearly one and a half centuries since their discovery. Only within the past few years have we begun to appreciate that these support cells serve critical functions in brain health and disease. Support cells are coming out of the closet.

A thumbnail history of support cells enables one to appreciate how these brain phantoms emerged from obscurity and moved to center stage.

The historical string of anecdotes also represents a case study of the circuitous, often confused labyrinth of scientific progress. The characteristic stages of discovery often can be captured by typical questions: "Are these cells new?" "Are they really different?" "How many types are there?" "What in the world do they do?"

The great German neuropathologist, Rudolph Carl Virchow, first described the brain's connective tissue in 1846.[1] This tissue appeared between the neurons, perhaps supporting them and holding them together. He used the word, *Nervenkitt*, to describe this new element. Many scientists translated the term *Nervenkitt* to mean "nerve-glue" or "nerve-cement." But, as the German neuroscientist George Somjen pointed out, the appropriate words would have been *Liem, Klebstoff,* or *Zement*, to denote glue or cement.[2] What, then, did Virchow have in mind? In Virchow's mind-set of mid–nineteenth-century Germany, he envisioned a sticky stuff that had bulk and shape. It was not thought to be stone-hard, however. "Putty" appears to be the nearest English equivalent. Alas, the terms "neuroputty," "brain putty," or "puttycells" never ignited scientific fancy. *Nervenkitt* was translated as "neuroglia," which captures the low esteem in which the support stuff was held. "Glia" is derived from the Greek word meaning "glue."

Though incorporating demeaning misconceptions, "glia" stuck like a cursed nickname. Since the term "glia" is uniformly used and recognized in neuroscience, I'll conform (under protest). Henceforth, I use the term for convenience, instead of the cumbersome "non–neuronal support, cement-like stuff."

A little more history before we plunge into the revolutions of modern glial biology. As best we can discern through the distorting lens of time, the renowned neuroanatomist, Deiters, in 1865, was the first to identify brain cells that were distinct from neurons.[3] These presumed connective tissue cells, now known as "neuroglial cells," or "glia" for short, were called "Deiters' cells" for most of the nineteenth century. Peering through the microscope, Deiters thought these cells could not be neurons because they lacked the long fiber that connects them to neurons: the axon (known affectionately in German by the cuddly name *Hauptaxencylinderfortatz*, a word which is a few millimeters longer than most axons). Deiters' drawings survive. Most neuroscientists examining them today cannot be sure whether Deiters had discovered a new cell group, or whether he stumbled upon yet other groups of unusual neurons with distinctive features. Regardless, a nineteenth-century cottage industry founded on glial cells flourished, rescuing hordes of pathologists and neuroanatomists from unemployment.

From plow to stirrups to airplane, progress of our fair species often depends on new tools. In the late nineteenth century the famed anatomist Camillo Golgi developed an astounding technique for staining cells in the brain. Unlike earlier methods, the Golgi technique stained cells in their entirety, allowing the microscopist to see every individual fiber of the stained cells. Golgi confirmed Deiters' contention that of all criteria proposed, only the absence of the long axon fiber distinguishes glial cells from neurons. His revolutionary technique and his careful observations clearly defined glial cells as separate, and moved the study of glia into the modern era. The battle over whether glia are truly separate entities came to a close; glia could be studied openly without inviting opprobrium.

The Glial Family

The field could now move to the next question: "How many types of glia are there?" Answers were critical if functions of these new glue cells were ever to be discovered. Scientists followed their inborn drive to discover, name, sort, and classify. The participants included many of the architects of modern neuroscience. The great Spanish neuroscientist and future Nobel Laureate, Santiago Ramón y Cajal, recognized one large family of star-shaped (aster) glia and appended the name "astrocyte" to this group. He published his work in 1909 and 1913.[4] The astrocytes were everywhere in the brain and spinal cord. They were in the gray matter, which is packed with nerve cell bodies, and in the white matter, which contains nerve fibers.

Cajal's protégé, del Rio Hortega, discovered a new type of glial cell in 1920 and named it "microglia."[5] He prophetically regarded the new cells as inflammatory agents that protect neurons from invaders. In writing a legend for one of his scientific illustrations he described the microglia in Andalusian romantic fashion: "the brain's nerve cells have bodyguards which extend their tentacles in every direction, and hold back whatever might be noxious." He examined a brain afflicted with encephalitis; and under the microscope he visualized microglia that "resemble voracious monsters and are valuable assistants in cleaning the tissue of whatever has damaged the nerve cells." Like his mentor, del Rio

Hortega was blessed with the ability to view cells forever frozen in time under the microscope. From but a single frame he accurately conjured the mortal battle. Over the next century, the roles of the microglial body-guards for good and evil would emerge with stunning clarity.

With the identification of a new third relative, the glial family was complete. The discovery of this remaining glial type, however, was wrapped in controversy and mystery. Cajal himself had identified some ill-defined cells as distinct in 1913 and he termed them the "tercer elemento," third element, of the nervous system. Their identity was unclear. Del Rio Hortega continued the quest but in so doing he forever alienated his teacher. Del Rio Hortega localized the new glial type to brain white matter, which contains large fiber bundles connecting one group of neurons with others. He presented his findings, but Cajal never accepted the new glia as separate and distinct from the already discovered microglia. The disagreement escalated into a passionate dispute; constant debates ultimately ruptured the relationship between master and disciple. The new cells, oligodendroglia, survived but the friendship did not. A visiting medical scientist, Wilder Penfield, a man destined to become a legendary neurosurgeon in Canada, studied the new glia and discovered that the oligodendroglia insulated nerve fibers which conducted impulses through the white matter.[6]

Glial Job Descriptions

The completion of the glial family tree, and agreement that glia and neurons are different, invited speculation about what they do. The great Ramón y Cajal posed the question: What is the function of glia? His answer: Nobody knows. Nevertheless, creative conjecture had begun.

In retrospect, each suggestion imaginatively synthesized fanciful flight and fruitful fact. Golgi suggested that glia provide nutrition for neurons. Since glial cells frequently send one set of fibers to blood vessels and another to neurons, he surmised that glia transfer nutrients from blood to neuron. Golgi tacitly perpetuated the courtly model of servile glia nourishing the "noble elements," the neurons.

Cajal was unimpressed by Golgi's theory of glia as waiters and wine-bearers; he rejected the nutritive hypothesis. He also rejected the then-popular "glia as space-fillers and form-givers" notion. It is ironic that, although he nixed the separate existence of oligodendroglia pro-

posed by del Rio Hortega, he favored a function for glia that future discoveries ascribed to these cells. Cajal cited with approval his own
brother's contention that glia insulate nerve fibers and expedite the
conduction of electrical impulses. A few short years later del Rio
Hortega and Penfield demonstrated that oligodendroglia served this
function.

Before the close of the nineteenth century, another function was
ascribed to glia; they engulf (called "phagocytosis," or cell-eating) and remove dying neurons. Penfield confirmed these observations during his
work in Cajal's institute in Madrid in 1924. He identified del Rio
Hortega's "bodyguards," the microglia, as the engulfing glial subtype.
Penfield described microglia as scavengers wandering through the brain,
cleaning up debris.

In one final discovery, in 1910, investigators detected granules in
the glia that resemble granules that store and secrete biochemicals in
other cells. Are glia actually endocrine-like cells that secrete stuff into the
bloodstream? Cajal greeted the notion with favor, concurring that
astrocytes are secretory cells. Others agreed, speculating that glia secrete
signals into the bloodstream and help the brain to communicate with peripheral organs, including the heart, liver, and kidneys.

The Glial Prophet

Ramón y Cajal, Golgi, del Rio Hortega, Penfield—four prophets who
helped create brain science. Their names are known to all neuroscientists.
But perhaps the most astounding essay on glia was written in 1907 by a
long-forgotten scientist. Professor E. Lugaro was director of the presti-
gious Institute of Mental and Nervous Illnesses in Messina, Italy. Though
respected and referred to extensively by his colleagues, he is a prophet
without honor among his scientific progeny. Yet in one shining scientific
publication he introduced a world of original ideas that set the scientific
agenda for the next century.[7]

He began with a critical analysis of prevailing concepts of glial function. He summarized the conventional wisdom of the time that glia were
"a sort of connective tissue," that served to "fill in the gaps" between
neurons, and that "they were little more than packing material." He ap-
provingly cited Cajal's logical objection to the contention "that the
neuroglia creates space for itself of its own volition" for physiological

purposes. Lugaro generalized his criticism of the faulty reasoning by noting that "it could be said that every object fills an empty space since it occupies a space that is not occupied by other objects."

He then proposed an astonishingly fertile array of hypotheses, some a century ahead of their time. Lugaro maintained, for example, that glia guide embryonic neurons during developmental migration; conclusive evidence for this wasn't found for another 50 years. And he further suggested a guidance mechanism: glia guide neurons by "chemotaxis," a mechanism that uses chemical cues and signals to attract and steer neurons on their migratory routes. Chemotaxis is only now being elucidated, as we enter the twenty-first century.

He next turned his attention to astrocytes, Cajal's star cells, in the adult. His overarching concept was that glia maintain a healthy environment for neurons. By describing component mechanisms, he formulated another scientific program for the next century. He postulated that glia adjacent to blood vessels prevent toxic substances from entering the brain, which anticipated the idea of a blood-brain barrier, currently a hot topic. We now know that this barrier also prevents life-saving medicines, as well as toxins, from entering the brain; an industry has developed to selectively breach this fortress, delivering treatments to the diseased brain. He also proposed that glia remove toxic waste products elaborated by neurons, and in so doing maintain the integrity of the neuronal environment. Scientists are now compiling a list of neuronal products taken up by glia; many of these chemicals are lethal for neurons.

In a baffling flight of scientific intuition, Lugaro turned to the glia's role in neuronal communication. He noted that glia surround and invest "nervous articulations," now known as synapses, the communicative junctions between nerve cells. He opined that these glia take up the signals through which one neuron excites another; hence glia can regulate neuronal communication by influencing the signals. In one leap, Lugaro anticipated roles for glia in synaptic communication, the discovery of neurotransmitter signals, the need for termination *and* initiation of excitation, and neuronal-glial-neuronal information flow. Lugaro had placed glia at the center of synaptic function, potentially at the center of learning and memory, and derangements thereof, provisionally including dementia. The role of glial dysfunction in Enoch's deterioration is very much an open question. As we enter the twenty-first century, we have not quite caught up with the wisdom of this forgotten prophet.

The fertile flurry of activity, deduction, and speculation at the turn of the century provided a ragged pedigree of glial family members and educated guesses about their functions. There are three main types of glia. Astrocytes maintain the neuronal environment, can provide neurons with trophic support, and play critical roles in neuronal communication. Oligodendrocytes, principally in the white matter, insulate nerve fibers, allowing efficient conduction of electrical impulses. Microglia are immune-inflammatory bodyguards, protecting against invaders and engulfing debris from the battles of illness and injury. How does this account accord with current views?

Glia, Growth, and Regeneration

By following the roles of glia in one prototypical function, the development and maintenance of information flow, we may begin to understand how glial dysfunction can result in disorders such as Enoch's dementia. During development glia form highways, complete with chemical signposts, and attractive and repulsive traffic signals that guide migrating neurons to their proper locations. Glia thereby generate the geography of brain communication.

The growth of fibers between proximate and distant neurons, allowing effective communication, is also guided by glial chemical signals. Both attractive and repulsive glial signals guide the growing fibers to targets, allowing appropriate synapses to form. Repulsive signals prevent growing fibers from wandering down blind alleys or forming improper connections, resulting in short-circuits and miscommunication. (However, attractive and repulsive signals may spell danger [and opportunity] later in life. Recent work, for example, suggests that elaboration of repulsive glial chemicals after injury prevents regrowth in the damaged spinal cord. Experimental inhibition of repulsive signals is now fostering regrowth. On the other hand, attractive signals are now being used as experimental treatments after illness and injury.) In sum, glia play key roles in the formation of functional communicative synapses. Although this scheme is more complex than originally conceived, Lugaro would have been proud.

Synapse formation itself is a multistep process that is orchestrated by glia. Experiments on the role of glia in synapse formation in cell culture are particularly illustrative. In the absence of glia, neurons from the retina form structural synapses in culture, which can be detected under the microscope; however, the synapses do not function normally. The synapses exhibit little spontaneous electrical activity. When stimulated electrically, they show a high rate of failure of information transfer. In contrast, if glia are added, the number of individual electrical signals that the neurons send on their own increases seventyfold.[8] While the exact glial actions responsible for these synaptic effects remain to be defined, it is clear that glia are necessary for normal, functioning synapses. Abnormal glial regulation of synaptic communication may play an important role in Enoch's mental debility.

After guiding synapse formation, glia become critical for synapse function. Glia surround synapses in the brain. Almost a century after Lugaro, the pieces of the synaptic glial puzzle are finally beginning to cohere. Glia are in the right place, the synapse, to influence synaptic function. The glial cell surface is covered with receptors for synaptic transmitter signals that relay information from one neuron to another. By binding to the glial receptors, synaptic transmitters can regulate glial function. Which functions? The discovery that glia produce growth and trophic (survival) factor signals, to which neurons respond, raised one important possibility. Can synaptic transmitters, that convey information between neurons, also govern factor production by glia? The answer is a resounding "yes." The synaptic transmitters, glutamate and acetylcholine, released by the transmitting neuron to excite the receiving neuron, also act on the glia. These transmitters increase the production of nerve growth factor (NGF) and brain derived neurotrophic factor (BDNF; see Chapters 4 and 9).[9] Production is increased at the most fundamental level, that of gene action. Synaptic neurotransmitters activate the genes in the glial cells that code for NGF and BDNF, increasing synthesis of these survival factors.[9] In turn, NGF and BDNF act directly on neurons.

NGF, as an example, accelerates nerve fiber growth, a prerequisite for connectivity and synapse formation. NGF also raises the number of synapses. Current discoveries indicate that the NGF family of trophic factors also strengthens synapses. Through these actions, glia may govern communication at synapses throughout the brain. Thus glia may influence functions like learning and memory, and when deranged may result in mental deficits such as those afflicting Enoch. BDNF, in turn, critically regulates synaptic transmission (see Chapter 9).

In view of their critical roles in communication, it is hardly surprising that glia are primary actors in instances of brain injury and disease.

From Enoch's Alzheimer's to stroke to Parkinson's disease to spinal injury and paralysis, disrupted connectivity and communication lie at the heart of crippling devastation. We need look no further than Enoch Wallace to appreciate the dysfunction elicited by disconnection and deranged communication in the brain. By harnessing glial capabilities, we may reestablish neural connections and communication after illness and injury.

❖

Enoch returned to Dr. Rudick for his follow-up exam, having no memory of his previous visit. He performed poorly on tests of cognitive function, frequently even forgetting the task at hand. His obvious failures were a source of shame and frustration, eliciting painful laughter, tears, and, finally, a rage reaction. Memory itself was identified as a discrete faculty during the sixth century B.C. in Greece, and its study was advanced through the Roman and medieval eras. Millennia of groundwork defined loci of memory in the brain, systems involved, associated synaptic processes, and a number of underlying cellular and molecular mechanisms. Ironically, as Enoch deteriorates, an ever-clearer picture of the nature of memory and its disorders is emerging.

❖

8

THE BRAIN CAPTURES TIME

❖

ENOCH AND SALLY returned to the neurologist for a follow-up examination. At first Enoch could barely remember the initial interview and physical exam, but the office and its unique paraphernalia jogged his memory. The odd juxtaposition of reflex hammers, tuning forks, ophthalmoscopes, eye charts, visual field screens, pins for sensory testing, and brain models scattered about the office brought fragments of his experience back into consciousness. The memories were accompanied by a vague, attendant anxiety. Addressing Sally and Enoch, the doctor reported that the CAT scan and MRI were normal, though he did not seem to attach much significance to the results. Enoch carried his sense of unease into the examining room.

Dr. Rudick looked vaguely familiar in his long white coat. The blue-topped hatpin in his lapel, the tuning fork in his left hip pocket, the handle of the reflex hammer protruding from his right pocket, and the stethoscope around his neck helped to orient Enoch to his present neurological reality. The doctor sounded far away as he talked about reevaluation, and said confusing things like, "a global impression of change," . . . "activities of daily living," . . . "and a self-maintenance scale."

Enoch wasn't sure whether these questions were part of a test or whether Dr. Rudick was explaining something. Enoch was becoming agitated. His mouth was dry; he had trouble swallowing. He felt his heart beating or skipping beats. Perspiration ran down his back, under the thin, blue gown. He was short of breath. He experienced an oppressive but ill-defined pressure in his chest. He was nauseated; he thought he would vomit. Although he was light-headed, he couldn't sit still. He squirmed in his chair, got up, paced across the tiny examining room, and sat down again, feeling even more uncomfortable.

"Mr. Wallace, are you OK?"

"I'm just fine, thank you."

"Would you like some water?"

"No, thank you."

"How have you been since your last visit?"

"Fine, thank you."

"I would like to ask you some questions."

"Yes . . ."

"I want you to remember some words."

"OK."

"Remember the color red, the number 32, and the name George Washington."

"OK."

Dr. Rudick then asked Enoch to name the fingers he held up. Enoch was more confused than wrong. He couldn't quite think of "index finger," "thumb," or "pinky," and used the word "band," instead of "ring" for the other finger.

Though the words were on the tip of his tongue, he couldn't name a pen, a book, a baseball, or glasses. Yet when shown a fountain pen, he correctly chose the word "pen," and not "book," "baseball," or "glasses." Enoch easily made the correct choice each time.

Dr. Rudick asked Enoch to name the words he had been asked to remember. Enoch stared at him blankly.

"I asked you to remember three words."

Enoch wished Dr. Rudick would just go away.

"Do you remember any of the words?"

"Yes."

"Which words do you remember?"

"I remember the pen."

"I asked you to remember three words."

Enoch was genuinely confused. He remembered naming some objects. He remembered trying to name fingers. He was not sure whether Dr. Rudick was referring to something else.

"Mr. Wallace, I'm going to ask you to follow some directions."

"OK."

"Close your eyes, stick out your tongue, and put your right hand on your head."

Enoch followed the orders without hesitation or error.

"Good. Raise your left hand, sit in the chair, and cross your legs." Enoch moved through the commands effortlessly. There were no signs of confusion. Enoch grew more confident. It took a while to get used to the strange proceedings of a neurologic exam. Anyone would be confused at the outset, but that wasn't a sign of abnormality. For the first time in a long while, he began to feel like himself—like the sharp Enoch Wallace.

"Good. Mr. Wallace, I would like you to perform some other tasks as I tell them to you."

"Fine." Enoch was now brimming with confidence. He felt alive.

"Wave goodbye."

This was so simple, it was almost silly. Enoch waved.

"Shake hands."

Enoch shook Dr. Rudick's hand. He was nearly euphoric.

"Light the match."

Enoch took the book of matches, deftly removed one, closed the book and lit the match carefully before blowing it out. Enoch was so happy and self-satisfied that he began laughing. Tears streamed down his cheeks.

"Why are you laughing, Mr. Wallace?"

It seemed so obvious, but Enoch had no easy answer. He was not quite sure why he was laughing. He was not even aware that he had been laughing. Enoch began to experience the empty, mild confusion again. Suddenly he felt like crying, but he caught himself. As Enoch moved precipitously from euphoria to dread, Dr. Rudick studied him from a distance. Enoch felt like a specimen on exhibit. He definitely did not like Dr. Rudick. The neurologist was cold and uncaring. Enoch was sure that he was just another peculiar case report to the clinician. Dr. Rudick cared about his reflex hammers, his charts, and his fees; he did not care about Enoch Wallace. Enoch was angry with Dr. Rudick. What right did he have asking Enoch to perform all these embarrassing tasks? What is so important about waving goodbye? Who cares about lighting a match?

"The name George Washington," burst from Enoch's lips in anger.

"What?" Dr. Rudick now seemed confused.

"George Washington. You asked me to say George Washington. You give so many orders. No one can keep them all straight. I remember George Washington. I'm not crazy. George Washington. I'm not sick."

Dr. Rudick matter-of-factly noted the outburst and proceeded.

"Mr. Wallace, I'd like you to draw some objects for me. Could you draw a clock that says 6 o'clock."

Enoch took the pencil and paper with a sense of hopelessness. He knew this was not going to turn out well. The circular outline of the clock's face looked more like an inverted *U*. The line was not even continuous. Enoch knew it wasn't right but didn't know how to correct the errors. And the numbers. The numbers were all over the page. The numbers were on the clock's face and off the clock's face. It looked all wrong, but Enoch was at a total loss when he tried to draw the clock correctly. And now the hands. What time did the doctor say? Enoch panicked when he couldn't remember the time.

"What time?"

"What?" Dr. Rudick again seemed confused.

"What time? What time did you say?"

"Oh. 6 o'clock. I said 6 o'clock."

Enoch made a complete mess of the hands of the clock. The hands did not stay on the clock's face. And since the numbers were scattered across the page, Enoch was confused about directions and angles and positions of the hands. In his frustration and confusion with the entire neurologic exam, and his anger at the demanding Dr. Rudick, Enoch suddenly tore the paper and threw it on the floor.

"I've had enough," he screamed at Dr. Rudick.

Far from being discomforted, the neurologist seemed to regard the outburst as routine. This annoyed Enoch even more. But Dr. Rudick persisted.

"Could you address this envelope to your wife?"

Enoch took the pen and envelope, but wasn't sure how to proceed. He flattened the envelope against the tabletop and placed the penpoint against the paper, but he didn't know how to begin. He forgot the task at hand. He searched Dr. Rudick for a clue, a prompt. The neurologist just studied him in silence. Enoch started writing his own name on the envelope but became distracted, and the line of characters trailed off. Enoch put the pen on the tabletop and stared off blankly with unfathomable pain and hopelessness. He now was beyond fear and crying and anger. He was alone.

Though Enoch is suffering from multiple, complex deficits, memory derangements are increasingly prominent. To focus on memory, we introduce another patient, who exhibits a dramatic and bizarre memory disorder.

❖

Elias Meharry was transported by police ambulance to the Accident Floor of the Boston City Hospital at 1:20 A.M. on a frigid, still, January night. The police found him nearly unresponsive in the abandoned school bus that he called home. His bloody left eye was swollen shut, frozen mucous stretched from nostrils to mouth, his layers of filthy shirts, sweatshirts, jackets, and outer pea coat were saturated with vomitus. His pale yellow jaundiced skin reflected the alcoholic's cirrhotic liver. The entire bus had been putrefied with the sweet-sour smell of whiskey.

On the ambulance litter, the stuporous Elias began stirring, perhaps responding to the warmth. His movement seemed to make the stench worse. Alcohol, vomitus, blood, and urine saturated the once antiseptic ambulance. He began tossing, rolling, and groaning. Reaching the hospital twenty minutes later, he was mumbling a few words, his eyes were open, though he was hardly alert.

As the Accident Floor doors swung open automatically, the medical team leaped into action for one of their regulars, Elias Meharry. Within five minutes, he was on a clean hospital litter, in a gown, IV fluids started. The sixty-second emergency exam had already begun. The obvious alcoholic received thiamine immediately, and anticonvulsants were ready at the bedside for rum fits. Medicine for "the DTs" (delirium tremens) were at hand. Elias was mumbling and confused, moving from apparent drowsiness to quasi-sentience. His eye movements were characteristically restricted, and jumped in the classical jerky (nystagmoid) dance. He remained on the overnight ward for close observation.

Though already battle-hardened world experts, no young house officer at the "Boston City" took the DT's casually. The psychotic, hallucinating, screaming, often violent patients were desperately ill, and suffered a 20 percent mortality rate. TB, pneumococcal pneumonia, hemorrhage, cirrhosis, seizures, and trauma were routine. Somehow, Elias made it through the long night without rum fit seizures or any sign of impending DT's. In fact, he brightened as the hours passed and by morning exhibited a characteristic semi-alert listlessness, dropping back into stupor only occasionally. He was alert enough to be examined.

He was transferred to the cavernous Peabody ward, an open barn-like, nineteenth-century affair of forty beds. Each bed was separated from those adjacent by wired glass partitions, affording the illusion of separation and privacy. The beds were arrayed along the periphery of the grimy, rectangular room, against the barely translucent windows. The center of the ward was empty, somehow accentuating the fractured, dirty yellow, tiny-tiled floor. An Inglenook fireplace improbably occupied one end of the room, opposite the bathroom on this men's ward.

Elias embodied the brain deficits of the chronic, malnourished alcoholic. Eye movements were restricted and still exhibited the classical jumping nystagmus. During his alert intervals, Elias was able to stand with help, and walk for the examining house officer. The broad-based, staggering, uncoordinated (ataxic) gait of alcoholic cerebellar damage was all too apparent. Sensory testing revealed the alcoholic peripheral neuropathy, which affects the long nerves outside of the brain and spinal cord. But the most dramatic abnormality was a bizarre amnesia peculiar to severe alcoholism.

Norm Hansen, a second year resident-in-training from Wisconsin continued his admission examination of Elias. He had finished the "physical" part of the neurological exam, and was now evaluating the patient's mental status. He asked Elias whether he remembered his (the doctor's) name. Elias broke into a wide grin

"Geez. Oh yeah sure, your name is Johnson Mahooty."

"Have you seen me before?"

"Sure. You were at the movie the other night."

"Thanks. I have to leave for a minute. I'll be right back."

"Take your time. Take your time."

Norm walked to the end of the ward and entered the nurse's station tiny room, out of view. He immediately turned around, re-entered the ward, and walked back to Elias' bed.

"Mr. Meharry, have you ever seen me before?"

"Sure."

"Where?"

"Geez, you were at the game."

"When was that, Mr. Meharry?"

"Oh, must'a been about last month."

"Have you seen me after that?"

"Oh, yeah."

"Where?"

"Geez. Saw you at the party."

"Have we met today?"

"Oh yeah."

"Where?"

"In the Fens, in the park."

"Were you in the Fens today?"

"Oh, sure."

"When?"

"Just before."

"Mr. Meharry, I've got a joke to tell you."

"Boy, I love jokes."

Norm then repeated the joke about the lawyer in heaven that he had just told to Elias before walking to the nurse's station. At the punch line, Elias burst into gales of laughter, as before, suggesting that this was a new one.

"That's a good one Doc. I got to remember that one. Geez."

"You ever hear that one before?"

"Geez I've heard jokes. That's a good one."

Several times during the day, Norm made a point of visiting Elias and recounting the same joke. Each retelling elicited the same burst of laughter with no apparent recognition. Elias was repeatedly hearing the same joke for the first time. In addition, Elias reacted to Norm's frequent visits as if he had never met him before, though upon questioning, Elias always had a new story about previous encounters.

While Elias exhibited widespread cognitive deficits, from difficulty concentrating to derangements of abstraction to visual-motor incoordination, the foregoing strange syndrome was dominant. He had a profound deficit in immediate/recent memory and made up stories, often blatantly absurd, to fill in the memory holes. Elias suffered from the syndrome of *amnestic confabulation*, memory loss and lying, which is characteristic of the alcoholic's Korsakoff's psychosis. The mental deficits are attributable to the death of neurons deep in Elias' brain.

The brain and memory. Is memory a discrete function of the brain? Where in the brain is memory? How do neurons make memories? How is memory organized? How many types of memory are there? How do we lose memory? What is forgetting? Read on.

Man and Memory

In the second half of the sixth century B.C., Scopas, a nobleman of Thessaly, hosted a sumptuous banquet feast.[1] The poet Simonides of Ceos was in attendance and, as was customary, dedicated a lyric poem to honor the beneficence of his vainglorious host. The panegyric celebrated Scopas and contained a graceful coda venerating the gods Castor and Pollux. The petty and malignant Scopas addressed the poet. Making common cause with tyrants through the ages, Scopas sarcastically informed the poet Simonides that he would receive only half the agreed fee.

"The remainder can be obtained from Castor and Pollux, the twin gods to whom you devoted half the poem." Fate is thus so easily sealed.

Host and guests continued to feast, while Simonides fell silent, seemingly in the depths of thought. The poet was roused from his reveries by a messenger: two young men were outside and wished to speak to him. Simonides, still in thought, arose from the huge banquet table and made his way out of the hall.

Outside there was no one. Simonides was alone. As he stood in silence, he was assaulted by a shattering roar: the massive roof of the banquet hall collapsed amid an explosion of dust and shattered stone. Scopas and his guests were crushed beyond all recognition. The monumental roof pulverized the corpses into a hideous carnage of flesh, blood, and bones. The twin gods, the phantom visitors, Castor and Pollux, had adequately repaid Simonides for his respect and adoration by luring him from the building just in time.

Mourning relatives arrived to remove the bodies for burial, but no corpses could be identified. Simonides, however, remembered the location of each guest at the table and guided the living to their appropriate dead. Having noted the seating arrangement, and having located each guest in sequence in his mind's eye, Simonides was able to reconstruct the scene from memory.

Beyond the evening's tragedy, beyond the mystery of his salvation by Castor and Pollux, Simonides realized that he had discovered a deep secret of memory: the faculty of memory could be cultivated and trained like other human skills; memory could be perfected through the use of techniques within the grasp of man. With Simonides' realization, the art of mnemonics was born. For the first time in human affairs, memory became a subject in and of itself, worthy of study. As related by Cicero in *De Oratore*, Simonides "inferred that persons desiring to train this faculty (of memory) must select places and form mental images of the things they wish to remember and store those images in the places, so that the order of the places will preserve the order of the things, and the images of the things will denote the things themselves, and we shall employ the places and images respectively as a wax writing-tablet and the letters written on it."[2]

In the ancient age of rhetoric, without pen or paper for notes, one thousand years before moveable type and printed books, a man was at the mercy of his memory. A skilled memory was the sine qua non of a successful life. Simonides' science of memory was liberating. With a trained memory, a teacher could teach as a master, a hunter could hunt in a wood not visited for ten years, a cook could prepare meals long forgotten by others. Without a trained memory, life in ancient Greece was lived darkly, in an eternal present. Simonides was as Prometheus, bringing the gift of memory.

Simonides focused on the psychology of memory, in addition to stressing the importance of an ordered *sequence* of things in proper *locations*. He discovered that the sense of sight and *images* are of the utmost importance in remembering. Cicero notes: "It has been sagaciously discerned by Simonides . . . that the most complete pictures are formed in our minds . . . by the senses, but that the keenest of all our senses is the sense of sight, and that consequently, perceptions received by the ears or by reflection can be most easily retained if they are also conveyed to our minds by the mediation of the eyes."[3]

And so was born the technique of memorization through the association of memories with images; so was born an appreciation of eidetic memory and "photographic memory." In its broadest sense, Simonides discovered the psychology of memory.

Simonides formulated the rules for sequences of places or locations. Then, five hundred years later in Rome, the rules for images were articulated. A long-forgotten teacher wrote the classic, *Ad Herenium*, which instructs students in the five parts of rhetoric, one of which is *memoria*.[4] Though the author treats all five parts, *inventio, dispositio, elocutio, memoria,* and *pronuntiatio*, in an instructional textbook fashion, it is as *the* source book of memory that *Ad Herenium* has earned immortality. It is the reference for the science of memory in the Greek and Latin worlds; it was the vehicle for the transmission of the science of memory to medieval and Renaissance Europe.

Pursuing this five-hundred-year-old inquiry, the anonymous author asks why some images are such potent stimulators of memory, while others are ineffectual. How do we discover the rules governing images for memory? *Ad Herenium* instructs:

> Now nature herself teaches us what we should do. When we see in every day life things that are petty, ordinary, and banal, we generally fail to remember them, because the mind is not being stirred by anything novel or marvelous. But if we see or hear something exceptionally base, dishonorable,

unusual, great, unbelievable, or ridiculous, that we are likely to remember for a long time. Accordingly, things immediate to our eye or ear we commonly forget; incidents of our childhood we often remember best. Nor could this be so for any other reason than that ordinary things easily slip from memory while the striking and the novel stay longer in the mind. A sunrise, the sun's course, a sunset are marvelous to no one because they occur daily. But solar eclipses are a source of wonder because they occur seldom. . . .

We ought, then, to set up images of a kind that can adhere longest in memory. And we shall do so if we establish similitudes as striking as possible; if we set up images that are not many or vague but active; if we assign to them exceptional beauty or singular ugliness; if we ornament some of them, as with crowns or purple cloaks, so that the similitude may be more distinct to us; or if we somehow disfigure them, as by introducing one stained with blood or soiled with mud or smeared with red paint, so that its form is more striking, or by assigning certain comic effects to our images, for that too, will ensure our remembering them more readily.[4]

The unknown author had rediscovered the psychology of memory. By appreciating the centrality of emotion and cognition in learning and memory, the anonymous author of *Ad Herenium* was not only describing a useful mnemonic trick, he was peering into the deep structure of the mind where emotion and memory are inextricably entwined. He was foreshadowing a science to be born two thousand years in the future, one that would begin elucidating the interactions of mind and emotion in the genesis of memory. That science would also begin defining brain structures underlying learning, memory, and emotion. The classical art and science of memory would ultimately provide the framework for understanding the cognitive–emotional deficits in diseases yet to be defined: Alzheimer's, schizophrenia, autism.

Of more immediate importance to citizens of the classical world, the art of memory worked in everyday life. A trained memory was capable of astounding feats. Seneca, a teacher of rhetoric, repeated two thousand names in the order in which he had received them. In a class of two hundred students, each recited a line of poetry. Seneca repeated each line, beginning with the two-hundredth and going backward in order to the first.[5] St. Augustine, another rhetoric teacher, told of his friend Simplicius, who could recite all of Virgil, backwards.[6] These mnemonic gymnastics attest to the efficacy of the classical art, and the ability to train, improve, and perfect natural memory.

If Simonides invented the art of memory, Aristotle provided the first organizing framework. His systematic approach represented an early inquiry into the nature of information processing in the formation of memory. As in so many spheres of knowledge, the paths lead back to Aristotle.

Aristotle applied his general theory of knowledge to memory. The perceptions transmitted by the primary senses are not directly processed for memories. Rather, the perceptions are initially manipulated by the *imagination*. The resultant images are the raw material for memory. "The soul never thinks without a mental picture."[7] "It is possible to put things before our eyes just as those do who invent mnemonics and construct images."[8] Memory and imagination are closely related parts of the soul. Memory is a storehouse of mental images derived from past sense impressions. Memory time stamps images originating in past sensory perceptions and stores the images. Aristotle's formulation, as with so much of his philosophy, has an eerily modern ring. He introduced the first stepwise scheme for memory processing, a tradition that lasted for the next two thousand years. The Aristotelean paradigm leads from art to science, and from mind to underlying brain mechanisms.

Aristotle's fertile conception unwittingly set the stage for an experimental psychology of memory. If memory were localized with sense perceptions in the same part of the soul, it could not be exclusively human. Animals sense and remember, too; so the processing of memories must be similar in them. Though Aristotle did not make the explicit leap, psychologists two millennia in the future did. The study of animals has been central to understanding human learning and memory, and its disorders.

Aristotle continues by examining the relation between recollection and memory. The process of recollection requires searching through memory archives to locate the volume of interest. The search is guided by two mental principles: association and order.[9] Though he never used the word "association," subsequent authorities cite Aristotle as the source of the rules of association based on similarity, opposition, and contiguity.[10] Association is used in conjunction with order because the volumes in memory exhibit the order of the original sense impressions ("timestamp," in modern jargon).

Aristotle integrates the principles of order and association:

> It often happens that a man cannot recall at the moment, but can search for what he wants and find it. This occurs when a man initiates many impulses, until at last he initiates that which the object of his search will follow. For remembering really depends upon the potential existence of the stimulating cause. . . . But he must seize hold of the starting point. For this

reason some use places for the purposes of recollecting. The reason for this is that men pass rapidly from one step to the next; for instance from milk to white, from white to air, from air to damp; after which one recollects autumn, supposing that one is trying to recollect that season.[11]

While the particularities of the associations may be elusive for the modern, the identification of order and association as guiding principles in recall is astonishingly lucid. Aristotle had set yet another agenda for the science of memory.

Aristotle turned his attention to the variables governing the persistence of memory, likening the laying down of memories to the impression a signet ring makes on the wax seal or tablet. He recognized that the process is imperfect, especially in the young and the old; and attempted to explain the limitations of memory in terms we would now recognize as physiological or pathophysiological. Indeed, Aristotle was well aware of the devastating and dehumanizing effects of memory loss. "Some men in the presence of considerable stimulus have no memory owing to disease or age, just as if a stimulus or a seal were impressed on flowing water. With them the design makes no impression because they are worn down like old walls in buildings, or because of the hardness of that which is to receive the impression. For this reason the very young and the old have poor memories; they are in a state of flux, the young because of their growth, the old because of their decay. For similar reasons neither the very quick nor the very slow appear to have good memories; the former are moister than they should be, and the latter harder; with the former the picture has no permanence, with the latter it makes no impression."[12]

From Simonides to Aristotle, memory as a discrete, definable faculty emerged as central to thought and mind. Plato's disciples and followers took memory one step further. For them, memory was no less than proof of the divinity of the soul. According to Plato, memory was not about the imagination processing sense perceptions. Memory is the soul's recollection of the eternal, archetypal ideas of the Truth, known before descent to this lower world. All knowledge is based on remembering the divine Ideas, of which earthly tokens are but pale reflections. Though differing radically in formulation and interpretation, Aristotle and Plato recognized that memory constituted a foundation of human mentality.

The neo-Platonist, Cicero, played a major role in the transmission of Greek rhetoric to Rome. He placed memory at the very center of the astounding faculties of man, associating *memoria* with *inventio*, which encompasses all human discovery and invention. In his meditations on memory, Cicero ponders the nature and origins of memory:

> For my part I wonder at memory in a still greater degree. For what is it that enables us to remember, or what character has it, or what is its origin? I am not inquiring into the powers of memory which, it is said, Simonides possessed, or Theodectes, or the powers of Cineas, whom Pyrrhus sent as ambassador to the Senate, or the powers in recent days of Charmadas, or Scepsius Metrodorus, who was lately alive, or the powers of our own Hortensius. *I am speaking of the average memory of man* [italics added], chiefly of those who are engaged in some higher branch of study and art, whose mental capacity it is hard to estimate, so much do they remember.[13]

Cicero's meditation continues, touching the human inventions of written language, astronomy, organized society, poetry, and philosophy. He conceives of *memoria* and *inventio* as the twin endowments of mind lying at the heart of these towering mental achievements: "A Power able to bring about such a number of important results is to my mind wholly divine. For what is the memory for things and words? What further is invention? Assuredly nothing can be apprehended even in God of greater value than this."[14]

The Christian Platonist and Latin Father of the Church, St. Augustine, saw memory as one of the three pillars of the soul: Memory, Understanding, and Will comprise the representation of the trinity in man. For Augustine, memory was one of the crowning achievements of man, the substance of self and humanity.

Augustine's awe is palpable before the palaces of memory containing the mental images derived from the senses. He is transported by the process of recall, which he describes with passionate immediacy. He examines the substance of memory, comparing sensory perceptions, memory of the arts, and memory of purely mental functions:

> Behold in the plains, and caves, and caverns of my memory, innumerable and innumerably full of innumerable kinds of things, either as images, as all bodies; or by actual presence, as the arts; or by certain notions and impressions, as the affections of the mind, which, even when the mind doth not feel, the memory retaineth, while yet whatsoever is in the memory is also in the mind . . . over all these do I run, I fly; I dive on this side and that, as far as I can, and there is no end.[15]

For Augustine, our memory, our minds render us human, in the image of God.

Where in the Brain Is Memory?

Even in the twentieth century, the golden age of x-ray, CAT scan, and electron microscope, our tentative grasp of brain and mind owes more to carefully observing nature in the manner of the ancients than to experimental wizardry. Brain illness and injury, those tragic experiments of nature, have suggested how the brain produces memory. Recognizing that the "very young and the old have poor memories," Aristotle irrevocably tied memory to biology. His observation that some have "no memory owing to disease" tacitly suggested that amnesia might be fruitfully studied in the ill. This challenge was approached in earnest two thousand years later far from the ancient world, in Montreal.

The Montreal Neurological Institute (MNI) is a mid-twentieth-century monument to the study of the brain and its disorders. Leaders from varied disciplines gathered at the MNI to plumb the secrets of the brain. Wilder Penfield, who pioneered the study of glia with del Rio Hortega twenty-five years earlier at Madrid's Cajal Institute (Chapter 7), emerged as a neurosurgical immortal, defining regional brain function on the operating table, at the MNI. In a parallel discipline, Donald Hebb, the famed neuropsychologist, propounded cell-based theories of learning and memory that still guide research today. Students flocked to the MNI from all corners. Discoveries, insights, and new approaches issued from the Institute monthly. In this charged environment, a group of psychologists and physicians turned their attention to patients with brain disease, hoping to understand the mysteries of memory.

Careful examination of patients by psychologists and physicians at the MNI had linked one particular area of the brain to memory. They could see that damage to part of the temporal lobe, on the side of the brain inside the temple, affected memory (Figure 8-1). Rather mild, highly specific memory deficits occurred with damage to the right or left temporal lobe. Patients with damage to the left temporal lobe, on the dominant side of the brain that produces language, experienced impaired verbal memory.[16] They were unable to remember words, whether they were heard or read. Surprisingly, however, their basic intelligence was unaffected.

Right, nondominant, temporal lobe damage presented a different clinical picture. These patients had entirely distinct memory deficits. They could not recall complicated visual pictures or complex sound pat-

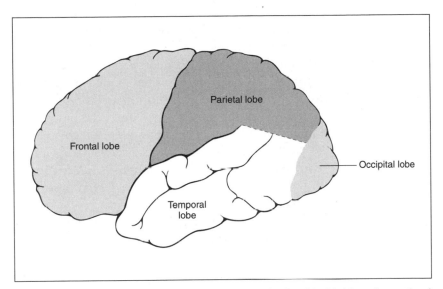

Figure 8-1 Lobes of the cerebral hemispheres. In this highly schematized representation, we are viewing the lateral hemispheric surface.

terns.[17] They also had difficulty learning to navigate maze puzzles on paper.[18] Much to the researchers' surprise, verbal memory was completely normal in these patients. Once again, their intelligence was unaffected, aside from the memory defect.

Was damage to specific parts of the temporal lobes responsible for impaired memory? This question was uniquely approachable in these patients because the temporal lobe destruction was unusually well defined; it had been caused by physicians. In an effort to help patients suffering from debilitating psychoses involving near-total mental fragmentation, and often requiring locked-ward, strait-jacketed self-protection, neurosurgical treatment was attempted. The temporal lobe was removed in an effort to prevent devastating hallucinations, delusions, thought disorders, and negativity. Neurosurgical approaches were also employed as a last resort in a group of patients with intractable epilepsy, who might endure tens or even hundreds of cataclysmic convulsions each day without warning, turning life into a horror. The already abnormal right or left temporal lobe that generated the seizures in these patients was removed neurosurgically to control the epilepsy. Temporal lobectomy relieved the epilepsy, but devastated memory.

Microscopic examination by countless neuropathologists and neurologists identified the walnut-sized hippocampus buried deep in the temporal lobe on either side of the brain as the critical memory structure. Memory deficits occurred when the horse-shaped hippocampus was removed with the temporal lobectomy. The hippocampus, and adjacent areas in the left and right lobes, appeared to be necessary for normal memory. While the identification of the hippocampus began to localize memory function in the brain, the MNI results raised even more perplexing questions. In general, why were the memory deficits of such little consequence for the patients? Why did surgically induced amnesia interfere so little with their day-to-day living? It seemed remarkable that these patients retained the ability to recall events in their lives, even while they could not remember words or complex pictures. All of the MNI neurologists had cared for patients with thoroughly debilitating memory loss, and were at a loss to explain why these severe syndromes weren't represented in patients with hippocampal removal.

Tentative answers were about to be provided by Wilder Penfield, the neurosurgeon, and his collaborator, Brenda Milner, a rising star in psychology. In the 1950s Milner pioneered the study of patients with brain damage to understand the relation of brain and memory; in a broader sense, her work defined the relation of brain and mind. Milner and Penfield were examining a patient named P.B, a highly intelligent civil engineer suffering from seizures. Penfield performed a left temporal lobectomy in P.B. in two stages to help relieve his epilepsy. After the first stage, in which the hippocampus was not removed, P.B. had little if any memory problem. Five years later, a second operation was performed, and the hippocampus was excised. Although both operations were confined to one side of the brain, P.B's memory deficit was devastating. He could barely care for himself, so severe was his memory impairment.[19] Why was P.B. so incapacitated by the unilateral (one-sided) left hippocampectomy, when all other such patients simply suffered mild verbal amnesia?

The mystery was solved twelve years later when P.B. died of a massive blood clot in his lungs, a pulmonary embolus. Examination of his brain revealed something totally unsuspected: The hippocampus on the opposite, right side was severely shrunken and damaged. Thus, in addition to having his left hippocampus removed, P.B. had an abnormal right hippocampus. In essence, P.B. was left with no normal hippocampal function after his second operation, a condition associated with devastating amnesia. P.B. was only one patient, however. The effects of bilateral hippocampal damage would have to be confirmed by studying other patients. Intensive examination of a new, hallmark patient was about to confirm initial impressions, and to revolutionize our understanding of memory and the brain.

Breakthrough at the Bedside

The young man who became medical history worked as a motor-winder.[20] He had been having debilitating seizures since the age of sixteen. The seizures were of the worst type, lasting an eternity of minutes at a time. He would fall to the floor, gripped by violent, uncontrolled, rigid shaking of his arms and legs, and lose consciousness. His jaw clenched as he frothed at the mouth; he was in constant danger of suffocating on his tongue. He lost control of his bowels. The violence of the convulsions posed the continual threat of bruises and fractures. Although he was treated with huge doses of anticonvulsant medications, the number and severity of his seizures continued to increase. Eleven years after their onset, at age twenty-seven, he was no longer able to ply his trade as a motor-winder. Life was contracting for this desperate young man.

Dr. William Scoville was consulted for a treatment of last resort. On August 25, 1953, patient H.M. was subjected to the radical approach of removing the temporal lobes on both sides of his brain. The hope was to prevent the life-destroying seizures. The patient returned from surgery in the accustomed groggy state. He remained drowsy for the next few days and could not be evaluated neurologically. As H.M. gradually emerged from his postoperative stupor, a dramatic, devastating memory loss became apparent. H.M. was completely unable to lay down new memories. He was thoroughly incapable of remembering any ongoing events. He could not even recall the surroundings or details of the house in which he had lived for six years. He could not remember his neighbors' names, or recognize them if he met them in the street. Paradoxically, however, he could remember his early life in great detail. H.M. suffered from a "dense" orthograde amnesia: he could not record new memories. He lived in an eternal present, and in the distant past of his childhood. Surgery had rescued him from the Scylla of intractable epilepsy and dashed him against the Charybdis of overwhelming, intractable amnesia. His memory deficit has persisted unchanged for nearly half a century to the present.

Paradoxically, H.M.'s intelligence increased after surgery, from 104 in 1953 to 118 in 1964. This was probably due to the marked decrease in the frequency of seizures. H.M. clearly demonstrated that memory function and intelligence, at least in certain forms, are independent, and most probably are produced by different brain mechanisms.

H.M.'s personality was also unchanged by the shadow of amnesia. He was described by Milner as "placid," "like his father," and only occasionally given to outbursts of "irritability."[20] (These simple bedside observations, made without the benefit of hi-tech imaging or genomic analysis, argued strongly for autonomous faculties of mind localized to different brain areas and systems. The independence of different mental capacities was becoming known as the "modularity" of mind. The deceptively simple concept of modularity would ultimately lead to deep insights into mind—from how it evolved, to right brain–left brain distinctions, to characterization of the unconscious.)

Milner proceeded beyond these initial observations to define the amnesia in detail with formal psychological testing. In her words, "forgetting occurred the instant his focus of attention shifted."[20] But if he was not distracted by other events, H.M. was capable of remarkably sustained attention to an event or details. He could hold facts in immediate memory if he attended to the task with all his might. He retained the number 584 for at least fifteen minutes by continually employing complex mnemonic strategies worthy of Simonides. H.M. revealed how his mnemonic secret scheme worked:

"It's easy. You just remember 8. You see, 5, 8, and 4 add to 17. You remember 8, subtract it from 17 and it leaves 9. Divide 9 in half and you get 5 and 4, and there you are: 584. Easy."[21]

Milner noted that a minute later, H.M. was unable to recall the number or, for that matter, any of the details of his complex mnemonic strategy. He did not even know that he had been given a number to remember because, in the interim, the examiner had changed the subject. Apparently, the only way H.M. was able to hold on to new information was through vigilant, continual rehearsal. Milner's testing, and H.M.'s reporting, suggested that there are distinct stages in the memory process. H.M.'s deficit revealed a very early stage of immediate memory, in which new information is held in the brain, perhaps in both hippocampi, before it becomes "consolidated" into a longer-term store. During this early period, a moment's distraction or loss of vigilance for an instant results in loss of the information. The fragility of this early stage must characterize memory in all of us, though we mercifully are unaware of our initial tenuous grasp of a new reality.

Tragically, H.M. was aware that a vital piece of his memory was missing, though the exact nature of his loss would always elude him. Between testing periods, he frequently looked at the examiner and remarked uncertainly and anxiously: "Right now, I'm wondering. Have I done or said anything amiss? You see, at this moment everything looks clear to me, but what happened just before? That's what worries me. It's like waking from a dream; I just don't remember."[22]

This poignant insight captures a consciousness whose mind has partially disintegrated, and hints at the deeper terror accompanying amnesia.

The Temporal Architecture of Memory

H.M.'s partial amnesia, and his coping strategy, seemed to confirm what had long been suspected: Memory consists of different phases in time. Further insights into the structure of memory could clarify amnesia and might provide an initial view of the structure of mind. Several teams of psychologists proceeded to test the cooperative H.M., employing ingenious approaches to tease out the temporal architecture of memory.

H.M.'s need to rehearse, and his failure to remember over seconds when distracted, suggested that he had a defect in some sort of "immediate memory" store. The first task, then, was to ascertain whether such a store could be identified. Testing with words or numbers was problematic because H.M. was a clever rehearser, and could appear to be recalling when he was actually just holding on to the present. In her doctoral thesis, the graduate student Prisko used an experimental method employing an array of perceptual stimuli.[23] She presented pairs of "clicks, tones, light flashes, shades of red (or) nonsense patterns" separated by intervals. H.M. was asked to determine whether the pair of stimuli were the same or different. To answer correctly, he had to hold the first stimulus in mind to compare it to the second. By increasing the time delay between the stimuli, Prisko determined whether H.M. had an immediate memory storage problem. Normal people do very well at this task. In matching twelve pairs of stimuli, they usually make only one error, even with delays of sixty seconds from the first to the second stimulus of a pair.

The performance of H.M. was clear-cut. With no time delay between the first and second stimulus, he made one error in twelve trials, a perfectly normal response. Yet, as the delay grew longer, H.M.'s performance deteriorated profoundly. With a sixty-second delay between presentation of stimuli, his performance was no better than chance. This result was confirmed by other scientists who used a related technique. H.M. was shown a series of ellipses, and then, after different delays, was asked to choose one of eight ellipses that matched the sample. With no delay, he performed normally. What was dramatic, however, was that he made no correct choices when the delays exceeded five seconds.[24] In contrast, normal subjects, including nine- to twelve-year-old children, are perfectly accurate with delays of forty seconds or longer.

Careful testing began to map the temporospatial architecture of memory. H.M. could immediately register new information, but he could not normally store information beyond the immediate present. The testing revealed a short-term memory compartment that holds new information for a number of seconds. It became apparent that this compartment is critical for laying down new memories. By implication, then, there is also a longer-term memory function that somehow overlaps with the short-term store. In sum, a provisional map of memory processing would include the movement of information from an immediate register to a short-term store to an intermediate or longer-term store. H.M. was acting as guide to the landscape of memory. He could receive new information and manipulate it, but he had no short-term, "holding" memory store from which it could travel to his long-term register. Lobectomy prevented the flow of memory into the long-term store.

The short-term memory store was not a disembodied, idealized construct. Neurosurgical data placed it squarely in the hippocampi and surrounding brain areas. In contrast, the immediate register did not appear to require the hippocampi. The two-thousand-year-old promises of Simonides and Aristotle were being fulfilled.

Milner now wanted to approach a trickier question. If H.M. could somehow be guided past the short-term memory bottleneck, would he then be able to retain information permanently? Stated another way, are the stages of memory separate and distinct enough so deficits at an early stage do not destroy function at a later stage? Are different memory modules insulated from each other in the brain?

Milner came up with an ingenious approach for bypassing H.M.'s short-term memory defect. At the time, H.M. was attempting to solve a complex puzzle without much success. While normal subjects solved the puzzle after twenty trials, H.M. didn't improve at all after 215 trials over three days. On reflection, the team recognized that this was simply an impossible task because of H.M.'s deficit. The maze had twenty-eight choice points. H.M. could not conceivably hold anything approaching that amount of information in a memory register that lasts only seconds. So the task was reformulated as a tiny maze with few enough choices to be encompassed by immediate memory time. After monumental effort, H.M. performed error-free, though it took 155 practice runs even on the watered-down maze.

Having mastered the maze, would H.M. retain the ability? After one full week's delay, H.M. was asked to solve the maze again. He performed without error.[25] His performance on relearning the short maze *two years later* was even more astonishing. H.M. required only thirty-nine practice runs (as opposed to the original 155) to perform error-free. Although he had retained information in some form for two years, he characteristically had no recollection whatsoever of the previous training sessions. As a control, two more choice points were added to the short maze. The maze now exceeded H.M.'s short-term storage capacity, and he failed completely: the short-term memory deficit remained.

H.M. was revealing the structure of memory. He retained 75 percent of the maze performance that he had learned two years before. Although his hippocampal damage interfered with the acquisition of new information after the fragile period of immediate storage, it did not prevent longer-term retention (weeks to years). H.M. also demonstrated that he could learn without having any recollection of having learned. He could know without knowing that he knew. Modularity of mind, indeed.

Milner was now ready to go after the modularity of memory more directly. Instead of testing verbal learning, pair matching, or maze learning, H.M. was asked to "remember" something else entirely, a motor skill. He was trained to mirror draw.[26] In this task, H.M. was asked to draw a line around a five-pointed star, staying within the double lines (Figure 8-2). There was a devilish catch to this task. He could not look at his hand directly or look at the star on the paper. Rather, he could only observe his hand and the star reflected in a mirror. H.M. exhibited a normal learning curve over the three-day training period. At first, he had a hard time performing the task, but he improved on each successive day. Indeed, he began each new session at the level he had attained at the end of the previous session. H.M. learned, and did so at a normal rate. Throughout the training, though, he was completely unaware that he had ever performed the mirror writing task before.

Milner had just made a landmark discovery. Motor skill learning occurred even when verbal learning, pair matching, and maze learning

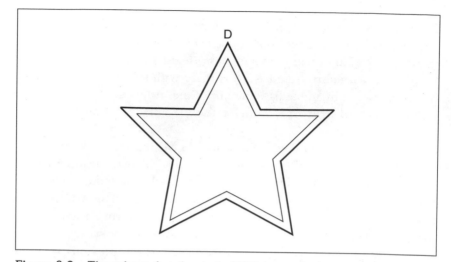

Figure 8-2   The mirror drawing task. H.M. was asked to trace the outline of the star with a pencil, remaining between the double lines. Crossing a line was scored as an error.

were devastated. The hippocampus may not be necessary for motor memory. Memory is modality-specific: The motor memory module must be distinct from the verbal memory module. Different brain systems must serve different memory modalities. In other words, the brain does not consist of an all-purpose memory system. Learning and memory are task-specific (or, more precisely, domain-specific).

Focusing on prototypical verbal memory, Milner and her collaborators had taken critical steps in describing the spatiotemporal architecture of the brain. Memory is not a unitary process; rather it is a series of overlapping processes. Immediate memory, lasting seconds, is distinct and can function in the absence of the hippocampi. Yet the hippocampi are necessary for a "transition process, or process of consolidation"[26] where information is moved to a short-term, more permanent store. Milner showed that the hippocampi are not the site of the long-term store since H.M. did exhibit retention.

The Architecture of Multiple Memories

Guided by H.M., students of memory divide the field into two broad categories: declarative memory and skill memory. Declarative memory is the conscious recollection of events of our lives and facts about the world. Skills, in contrast, consist of learned motor abilities—mirror drawing, bike riding, or dancing—and don't require recall.

H.M.'s declarative memory was devastated by hippocampal loss, but, as Milner's brilliant experiments with maze and mirror drawing demonstrated, his skill memory was intact. How distinct are declarative and skill memories; are they served by different brain systems? To approach these questions, it would be necessary to find patients who complement H.M. Are there diseases that selectively damage skill memory and leave declarative memory undisturbed?

Patients with classical amnesia have the same pattern of declarative memory deficits as H.M. In general terms, patients with early Alzheimer's disease resemble H.M., though the picture can be complicated in this disorder because many brain systems may be deranged. In dramatic contrast, patients with Parkinson's disease have normal declarative memory but impaired skill memory.

It isn't news that performance of conscious, motor (muscle-based) acts is deranged in Parkinson's disease. This degenerative disorder strikes middle-aged and elderly men and women. Patients typically suffer from slowed movement, muscle rigidity, a characteristic shuffling walk, a mask-like face resulting from stiff facial muscles, and a slow tremor of arms, hands, and legs. Beyond these obvious deficits, bizarre behavioral-motor abnormalities hint at deeper mysteries.

A typical history is illustrative. A sixty-six-year-old, retired white male accountant has had Parkinson's disease for nine years. He is now so debilitated that he is confined to a bed and wheelchair in a nursing home. He is, however, capable of taking two or three precarious steps with the support of an attendant, once he is helped to the standing position. His immobility has worsened over the past two years. Tuesday morning, after the nurse helped feed him breakfast, the normal routine of the home is suddenly shattered. The alarm bell begins ringing, and the siren shrieks as an aide in the opposite room screams "fire." The patient, in a panic, leaps out of the wheelchair, runs twenty feet to the outside door, and vaults to the front lawn. Having reached safety outside, he falls to the grass, immobile and virtually paralyzed. Life-threatening emergency over, he reverts to his frozen, Parkinsonian state. This strange episode is as familiar to caretakers of Parkinson's patients as slowed movement and difficulty walking. What happened?

Though we hardly have an adequate explanation of this occurrence, we can begin describing it in terms that relate to skill memory. Parkinsonian patients have difficulty initiating and executing voluntary motor programs. The programs are learned motor skills that involve multiple steps, such as arising from a chair, standing erect, initiating walking, and then proceeding with a normal forward gait complete with arm swing. Nevertheless, in times of emergency, these complex programs can be executed involuntarily. Voluntary motor recall is impaired, but involuntary recall is functional. There seems to be a deficit in the voluntary *retrieval* of motor skills. The skill is in the brain somewhere, but is only accessible in a perceived emergency. Where is the skill stored, and where might the deficits lie?

Patients with Parkinson's disease have abnormal nondeclarative, skill memory, but relatively normal declarative memory. Moreover, the main system that degenerates in the disease has been known for decades. The circuit is called the nigrostriatal system. Fibers run from the substantia nigra (black substance) in the brainstem at the base of the brain, up to the striatum deep in the cerebral hemispheres (see Figure 3-2 on page 58). Progressive death of nigral neurons has been shown to result in the debilitating motor symptoms of Parkinson's disease, though the cause of nigral cell death is unknown. When the nigrostriatal system is destroyed in experimental animals, from rats to monkeys, the symptoms of Parkinson's disease inevitably appear. Thus we can conclude with confidence that abnormalities of this system are responsible for the disease, and for the consequent deficits in skill memory.

Death of the nigral neurons deprives the striatal neurons of their normal input, which results in the motor deficits of the disease. Treatment with L-DOPA, a precursor of the neurotransmitter dopamine, partially replaces this nigral transmitter input to the striatum, and reverses many of the deficits of Parkinson's. The striatum itself is thought to be critical for motor memory. In turn, normal striatal function depends on input from the nigral neurons. So, striatal neurons, receiving critical input from nigral neurons, are necessary for skill memory. Conversely, hippocampal neurons are critical for declarative memory. Though we know little about striatal memory, we do know that this is the brain area to examine. We are more advanced in our understanding of hippocampal function in declarative memory.

Peering into the Hippocampal Black Box

Appreciating the existence of multiple memories, we return to the declarative memory as a prototype. Memory happens in the hippocampus. What does the hippocampus actually do? What near-magical transformations occur there? Can we discover the mystical (or mythical) memory "trace," the elusive memory "engram?" This quest assumed the epic proportions of the search for the missing link, the lost chord, the Northwest Passage, the Fountain of Youth, and the Holy Grail, all rolled into one. Of course, there was always the dreaded, half-articulated fear that the hippocampus contained a complexity of neurons, circuits, and mechanisms that would be thoroughly unrevealing when analyzed. The nihilists lined up against the true believers.

One roadmap was provided by the Montreal psychologist, Donald Hebb, in his pioneering 1949 book, *The Organization of Behavior.*[27] Considering the brain basis of learning, Hebb concentrated not on neurons themselves, but on the connections between neurons. Hebb speculated that learning consisted of strengthening the communicative connections, the synapses, between neurons. Strengthened connections bind neurons together into "ensembles." If individual neurons represent different features of a scene, for example, the ensemble recreates an image of the scene in its entirety, like the pieces of a puzzle. Similarly, strengthened synapses form an ensemble associating the visual, auditory, emotional, and conceptual features of a memory that are encoded in the component neurons. Hebb's fertile speculations shifted the focus of the search for learning and memory mechanisms to synaptic communication.

Hebb suggested that the more a synapse is used, the stronger and more efficient at transferring information it becomes. Use, in turn, is driven by experience. Our experiences use specific circuits and synapses, thereby strengthening those very synapses. This was a pleasing, parsimonious mechanism. Use itself recorded an experience in the synapses and pathways, resulting in learning and memory.

How does experience strengthen synapses? In Hebb's view, synaptic strengthening required the participation of both neurons forming the connection. Both the signal-sending, presynaptic neuron and the receiving, postsynaptic neuron have to fire (electrically discharge) to result in a stronger synapse. If the presynaptic neuron actively discharges and sends a signal, but the receiving neuron does not respond, strengthening does not occur. Or if the postsynaptic neuron discharges in the absence of presynaptic firing, that synapse is not strengthened. Synapses do not strengthen willy-nilly because scattered neurons fire; information must actually be transferred from sending to receiving neuron to ensure strengthening. This requirement implies that the neural circuit must be functionally active to result in synaptic strengthening and memory.

Hebb's elegant theory stood as a shining model, awaiting experimental verification from a real brain. It was a long wait. Nearly one quarter of a century after the publication of the 1949 book, the breakthrough occurred. There had been no dearth of experimentation. The hippocampus had become the focus of intense interest as the likely locus of memory. The anatomy of the hippocampus was being elucidated in detail, defining the circuits available to participate in memory mechanisms. In parallel, the biochemistry of the signals that the pathways use to communicate was gradually emerging with the advent of new techniques. Continued study of patient material was providing information about the hippocampus in disease, including stroke and epilepsy. The anatomical, microscopic, pathological, and biochemical studies provided important basic insights into hippocampal organization, allowing new approaches to the mysterious, memory mechanisms. The emerging blueprint of hippocampal circuitry encouraged electrophysiologists to study synaptic communication.

Nature has been charitable in the hippocampus; neural pathways are arranged in a simple fashion. They are discrete, well segregated, and conveniently identified. It is possible to use needle-like electrodes to stimulate one pathway electrically while recording from its target neurons without disturbing others. One circuit is particularly attractive for study. A large bundle of nerve fibers, that we will call the path, enters the hippocampus from the outside and forms synaptic connections with a specific group of target neurons. Electrical stimulation of the path excites the target cell fibers through the synapses and fires the cells themselves. Since the whole pathway is compartmentalized so well, this stimulation doesn't affect other hippocampal circuits. The path, the target cells, and the connecting synapses are so well organized that each element can be examined individually. It is as close to an ideal arrangement for study as the brain is likely to offer.

Examining the path-target circuit, the Norwegian electrophysiologist T. Lomo made a peculiar experimental finding in 1966.[28] While performing experiments on the hippocampus for other reasons, he delivered test shocks to the path and recorded electrical responses from the target neurons. Unexpectedly, a single test shock to the path magnified responses from the target neurons for ten to twenty seconds. In the lingo of hippocampal electrophysiology, the target responses were potentiated by the test shocks. Somehow the test shock enhanced subsequent communication of the path fibers with the target neurons. The result was unusual enough to be worth following up. Lomo joined forces with a British colleague, Tim Bliss, to pursue things.

Bliss and Lomo prepared rabbits by carefully implanting electrodes in their hippocampi and stimulating the path; then they recorded from different parts of the target neurons.[29] If an electrical shock really did enhance synaptic communication, they wanted to document it.

In a standard experimental protocol, they began by stimulating the path at the low rate of one impulse every two seconds. This was a control test to establish the baseline response of the target neurons. They followed the control period with the real test: the pathway was stimulated at ten to fifteen impulses per second for several seconds. They measured the size (amplitude) of the responding electrical potential in the receiving target cell fibers. Since the path makes synaptic contact with the target fibers, the magnitude of the electrical response of the target fibers is a direct measure of the effectiveness of synaptic communication. After several seconds, the stimulation rate was reduced to the initial low value of one impulse every two seconds, but the electrical *response* of the target fibers remained elevated. Bliss and Lomo found that the test stimulation of ten to fifteen per second caused a striking, long-lasting jump in the size of the target synaptic responses to subsequent low-level stimulation. After a brief increase in pathway stimulation, the heightened responses persisted for at least twenty minutes. Elevated stimulation for seconds increased synaptic responses to low level stimulation for at least minutes. How long would this heightened responsiveness persist?

Bliss and Lomo next performed a long experiment in which test shocks of ten to fifteen impulses per second for ten seconds were repeated four times over the course of two and one-half hours. They observed a threefold rise in target fiber synapse responses to subsequent normal, low-frequency stimulation. These responses persisted for at least six hours: increased electrical activity for seconds caused increased synaptic responses for at least hours.

Bliss and Lomo also measured responses in the target cell bodies. The cell bodies lie at a distance from the synaptic connections on the fibers. Synaptic transmission at the synapses creates an impulse in each fiber that travels to the cell body. At the cell body, the impulse causes another electrical discharge, called a spike. They measured the spikes after the test shocks and, sure enough, they enlarged after the test shocks; the increased responses also lasted for at least six hours.

Finally, they measured the speed of the spike response after stimulation, as another index of the effectiveness of path-target neuron communication. After the path was stimulated with elevated test shocks, the target cells responded far more quickly to baseline stimulation of the path fibers. The greater speed of response resulted from test shocks lasting seconds, but also persisted for at least six hours.

To summarize, increased stimulation for seconds results in dramatic strengthening of synaptic communication, neuron responsiveness, and speed of response that persist for hours after increased stimulation has stopped. An experience lasting about ten seconds increases responses for about six hours, or about 21,600 seconds [6 (hours) × 60 (minutes) × 60 (seconds) = 21,600 (seconds)]. This hippocampal phenomenon has some interesting properties.

The potentiation that Bliss and Lomo discovered had traits suspiciously reminiscent of memory. A brief experience caused a long-lasting change in synapses in the hippocampus, a locus of memory. Moreover, the presumed memory changes were profoundly important for brain function. Greater synaptic strength enhanced the neural pathway's function and accelerated information processing. In this dynamic view, memory did not consist of storage of indifferent tokens in a distant, dusty repository. Memory, incorporated by experience into synapses, altered brain function.

Bliss and Lomo also hinted at the answer to a long-standing riddle. How does the nervous system encode memories? What language is used to unambiguously store the smell of jasmine, the sound of Beethoven's Ninth, the vision of a wooded sunset, or the exhilaration of discovery? Their experiments suggested that experiences strengthen synapses of only the specific pathways processing an experience; experiences do not affect synapses that are not activated. Specificity is built into the process; memories are synapse-specific.

The synaptic model also explains the fidelity or precision of memory storage. Synaptic strengthening can be graded in that it permits, for example, a continuum of strengths to encode intensity and duration of experience. The activation of combinations of synapses allows nearly limitless richness of representation.

Enhanced synaptic strength, in turn, could result from changes in extrinsic influences, or from changes intrinsic to the synapse. Extrinsic influences include the effects of other nerves playing on the path-target system, that might augment responsiveness to excitatory extrinsic nerves or decrease the effects of inhibitory extrinsic nerves. The net result would be to enhance responsiveness of the synapses.

In general terms, intrinsic synaptic changes are restricted to two categories. Experience may potentiate the effectiveness of the signal sending side of the synapse or may enhance the responsiveness of the receiving side of the synapse. That exhausts the formal possibilities, though a multitude of complex, individual mechanisms might be involved in either alternative. Bliss and Lomo were unable to distinguish between these alternatives and, in fact, could not exclude the possibility that extrinsic factors were also important. As we enter the twenty-first century, mechanisms underlying synaptic strengthening continue to be the focus of intense research.

Bliss and Lomo concluded their groundbreaking article with a modest summary of the "significance of the effect":

> The interest of these results derives both from the prolonged duration of the effect, and from the fact that an identifiable cortical pathway is involved. The [perforant] path is one of the main extrinsic inputs to the hippocampal formation, a region of the brain which has been much discussed in connexion with learning and memory. . . . Our experiments show that there exists at least one group of synapses in the hippocampus whose efficiency is influenced by activity which may have occurred several hours previously—a time scale long enough to be potentially useful for information storage. Whether or not the intact animal makes use in real life of a property which has been revealed by synchronous repetitive volleys *[of impulses]* to a population of fibres the normal rate and pattern of activity along which are unknown, is another matter.[29]

The Combination Lock for Memory

Simonides brought us memory as a discrete phenomenon. Aristotle tied memory to biology. The Romans emphasized the centrality of emotion and cognition to memory, and St. Augustine placed memory within the trinity of human qualities. With the late-twentieth-century neuroscientific revolution, memory found a home in the hippocampus; particular neuronal circuits serving long-term potentiation (LTP), the reigning cellular model discovered by Bliss and Lomo, were identified. Continuing to zoom down, the synapse itself has been tentatively identi-

fied as the locus of memory formation. Are we now able to drive memory to the molecular level? At the cellular and molecular level of
understanding, we can begin contemplating specific mechanisms and
treatment of deficits.

In searching for candidate memory molecules, we must keep in
mind the peculiar functions that a memory molecule has to serve. LTP
remains our model. The strengthening of synaptic communication is
the gold standard; any molecule of interest should participate, if not
cause, strengthening. In turn, according to Hebb's rule, synaptic
strengthening occurs only when the presynaptic transmitting neuron
and the postsynaptic receiving neuron fire simultaneously, so random
firing of single neurons is prevented from forming meaningless memories. Candidate molecules must somehow record and react to this simultaneous firing.

We also have the matter of synapse specificity. Although each neuron forms and receives about 10,000 synapses, memories are probably
synapse-specific. Of these 10,000, only the few active synapses on the firing neuron should be strengthened. Active synapses form the circuits
stimulated by the experience to be remembered. Mediating molecules
must honor this extraordinary degree of specificity.

Candidate memory molecules must also mediate several circuit features that characterize LTP. For LTP induction, a firing intensity threshold must be reached in the memory neuron. This can be done by
intensely stimulating a single input fiber, or by stimulating *many* fibers at a
lower intensity. This phenomenon has been termed "cooperativity."
This characteristic is related to another trait of LTP, *associativity*. Synchronous low-intensity stimulation of many weak inputs converging on
the same neuron induces LTP even when none of the pathways alone is
sufficient. The input stimulation can occur across large distances on the
receiving fiber (dendrite), and derive from fibers arriving from widely
dispersed brain areas. Through associativity, independent experiences
could be bound together in memory. Weak input odorant fiber stimulation and weak visual fibers carrying the representation of a rose, could associate the smell and sight of the rose in memory.

What would a rose molecule look like? How could a single molecule possibly satisfy all the implied functions? What feat of biological engineering could design a molecule that senses that communicating
neurons are firing simultaneously, reacts only at the fraction of synapses
that are active in the neuronal firestorm, and detects cooperativity and as-

sociativity? How would such a molecule convert all this information into synaptic strengthening and memory?

One such molecule has been discovered. It is an ultra-special receptor lock into which the glutamate transmitter signal key fits.

Combination Locks and Coincidence Detection

Glutamate (abbreviated "glu") is the excitatory, amino acid transmitter that constitutes the LTP signal. In response to an electrical impulse, the transmitting neuron releases glutamate, which stimulates the receiving neuron at the synapse. Glu excites the receiving neuron by binding to receptors on its surface. There are many types of glu receptors on the receiving neuron, but only one particular glu receptor subtype is specialized for LTP. It incorporates all the features we described that allow it to function as a memory molecule. It is called the N-methyl-D-aspartate (NMDA) receptor.

As expected, activation of the NMDA receptor requires glu binding. A signal key activates a specific receptor lock in this standard operating procedure. However, the NMDA receptor incorporates a novel, decisive wrinkle designed for LTP and memory. Binding of glu alone does not activate the receptor. For activation, another condition must be met. After glu binding, the receptor is activated only if its own neuron fires electrically. Activation of this specialized receptor requires both glu binding *and* simultaneous firing of its own neuron. For this reason the receptor has been termed a "coincidence detector," activating only if glu binding and depolarization occur. The NMDA receptor consequently can decide whether Hebb's rule of pre- and postsynaptic firing are occurring together.

Timing itself represents another unique property of the NMDA receptor. For almost all other receptors, signal binding activates the receptor for 1 to 10 thousandths of a second (termed *milliseconds*, or *msec*). The standard NMDA receptor, in contrast, is activated for several hundred msec after glu binding. Consequently, the NMDA receptor itself has a relatively long memory. After a glu binding history, it remembers, and detects coincident pre- and postsynaptic firing occurring over the relatively long period of hundreds of msec. As a result, transmitting and receiving neuron firing do not have to be completely synchronous to activate the receptor. Flexibility is built into the coincidence detection mechanism.

NMDA receptors exhibit even greater temporal flexibility than I have described so far. Indeed, they are a family of receptors in which each member is made up of slightly different molecular building blocks. Though closely related, different building blocks change the timing of activation by glu and, consequently, the timing properties of coincidence detection: different family members have different memories. For example, one member has a memory of 100 milliseconds. Neuronal firing activates it for about one hundred milliseconds after glu binding. Another family member, composed of additional building blocks, remains responsive to firing of its own neuron for approximately four hundred milliseconds after glu binding. At the extreme lies a receptor that can be activated for 5000 milliseconds (five full seconds) after glu binding. These temporal characteristics allow different NMDA family members to serve different memory functions.

In very general terms, the developing embryonic brain is characterized by NMDA receptors with long memories. With maturity, receptors switch to shorter time sensitivity. The precision of neuron–to–neuron connections during the growth of nerve circuits is thought to depend on firing patterns. Neurons that "fire together wire together." NMDA receptors appear to participate in the formation of synaptic connections during development, as well as in the strengthening of synapses in the adult. The long-lasting sensitivity of receptors to neuronal firing after glu binding in the embryo may provide a safety factor for proper circuit formation, even if precise synchrony is not achieved.

The switch from long to short memory receptors may explain why critical periods occur during development. During a restricted period of early development only, chicks imprint on hens, humans, or other objects, and follow them dutifully as "mother." Visual deprivation during a restricted period of early infancy results in permanent blindness. Language learning is optimal during the first six years of life and decreases radically thereafter. The developmental critical periods may depend on the presence of "long memory" NMDA receptors.

Although conjectural at present, the time constants of NMDA receptor family members may meet the demands of different types of associativity in different brain systems. In some circuits incoming impulses may be dispersed over time, while in others near-simultaneity may be the rule. Some memory associations may require NMDA receptors with very different temporal characteristics. In contrast, systems designed to carry out complex associations among manifold incoming stimuli may require longer periods to become sensitive to neuronal firing.

Once activated, the NMDA receptors form a channel that allows electrically charged calcium (abbreviated Ca^{2+}) to enter the receiving neuron from outside the cell. Only activated receptor channels permit Ca^{2+} entry. Once inside the cell, Ca^{2+} acts as a signal that triggers multiple biochemical actions in the neuron: Calcium stimulates key enzymes, changes the shape of critical proteins, and turns on selected genes. All of these actions may alter synaptic strength, although the details are still being defined.

What about synaptic weakening? If experience and neural activity increase strength, does idleness decrease strength? More generally, do certain processes vary synaptic strength in the brain, either up or down? The answer is an emphatic "yes." LTP has its mirror image in LTD, long term depression. While synchronous pre- and postsynaptic discharge increase strength, *asynchronous* firing decreases strength. In sum, the same synapse undergoes graded weakening or strengthening according to experience. The brain continually monitors experience and adjusts accordingly. It emerges as an ever-growing, changing, living structure that learns from experience. The brain bears scarce resemblance to the immutable, hard-wired switchboard envisioned a few short years ago.

Synapse Strengthening Signals

Once the NMDA receptor detects coincident firing of the pre- and postsynaptic neuron, how is synaptic strengthening actually accomplished? Does the NMDA receptor somehow activate another molecule that actually strengthens the synapse? It is not surprising that signaling molecules are thought to play a critical role in synaptic strengthening, informing each neuron about the firing state of the other, causing the strengthening itself, and maintaining the strengthening over time. How would we recognize a candidate strengthening signal molecule if we found it? What would it look like?

A synaptic-strengthening signal molecule should fulfill several minimal criteria. The molecule should strengthen synapses. The strengthening should be long lasting. Experiences that increase synaptic activity or transmission (à la Bliss and Lomo) should increase the availability of the signal molecule, eliciting synaptic strengthening. Though not obligatory, the molecule might be conveniently located in the pre- and/or postsynaptic neurons, allowing immediate access to the active synapse.

How could we identify molecules with these characteristics? Do any known molecules fulfill these unusual criteria? One group of molecules exhibited at least some appropriate properties. Trophic or survival factors, discovered by Levi-Montalcini half a century before, during World War II, possessed some interesting traits (see Chapter 5). Though these factors were thought to play critical roles during development of the nervous system exclusively, their actions were potentially relevant to synaptic strengthening in general. Nerve growth factor (NGF), the prototype, elicited long-lasting changes in neuronal connectivity. NGF caused neurons to extend fibers to other neurons. Under the influence of NGF, the growing fibers formed synaptic connections and operational neural circuits that conducted impulses. These changes were certainly long lasting, persisting for days to weeks, at least. On the other hand, trophic factors did not appear to act quickly enough to serve synaptic strengthening, which occurs in seconds, as Bliss and Lomo showed. Nevertheless, trophic factors were long known to alter neuronal excitability, an important property of a synaptic signal. Could NMDA receptors strengthen synapses through the mediation of NGF-like molecules?

In the 1960s and 1970s, actions of NGF were still being catalogued. At the time, NGF was thought to act exclusively in the peripheral nervous system. The trophic factor was added to peripheral neurons in culture and its effects were studied. The commonly employed test cells were peripheral sensory or sympathetic neurons, or related tumor cells that were easily grown in culture. PC 12 cells were one such tumor, which was closely related to adrenal cells and sympathetic neurons. NGF dramatically increased the electrical excitability of the cells in culture. Several mechanisms underlying this striking effect were identified. NGF increased the number of electrically conducting (ion) channels in the nerve.[30] These submicroscopic channels allow electrically charged atoms to flow into and out of the firing nerve, generating the electrical impulse.

NGF also elicited a separate but complementary effect. The trophic factor increased responsiveness of neurons to neurotransmitter signals. Transmitters are the chemical signals (such as glu) that convey the electrical impulse from one neuron to another across the synaptic gap. NGF increased the number of transmitter receptors on neurons, resulting in enhanced responsiveness. So, NMDA receptor stimulation may result in activation of an NGF–like molecule that actually strengthens the synapse.

In sum, these discoveries indicated that NGF increased responsiveness to transmitter signals sent from other neurons, and increased generation of electrical impulses. At the time, 1980, NGF was regarded solely as a developmental survival agent; actions on neuronal excitability were thought to enhance development and differentiation into mature neurons. In 1980, NGF, the only known trophic factor, was not even thought to be in the brain. As always, chance favors the prepared. The discovery that NGF was but one member of a trophic factor family (Chapter 5), and the detection of trophic factors in the brain in the late 1980s, raised a host of new possibilities. We now know that members of the NGF trophic family can strengthen synapses in response to firing. Intense investigation is now devoted to unraveling mechanisms of action.

From Simonides to synapses, twenty centuries of inquiry have begun to yield the secrets of memory and brain function. In the process, we are gaining an awareness of the functional impairments destroying the life and mind of Enoch Wallace.

❖

Sally felt life slipping beyond control as Enoch demanded near-constant protective care; Dr. Rudick advised a changing lifestyle with the possible diagnosis of Alzheimer's disease. What had she done to bear witness to the destruction of Enoch's humanity? The intimate relationship of development and destruction had become apparent to neuroscientists only recently. The very same molecular mechanisms governing brain function, including learning and memory, elicited cell death when out of control. These insights have profound implications for stroke, brain trauma, and Alzheimer's itself.

❖

9

GROWTH THAT KILLS

❖

SALLY AWOKE suddenly. Her heart was racing. Thank God it was a dream. She and Enoch were walking on a rocky beach in Martha's Vineyard. Then he was in the middle of a steel jungle gym, surrounded by bars. She couldn't reach him. He was trapped in the middle of the maze. He was just a few feet away, but she couldn't reach him. She was paralyzed with dread. She could feel her heart beating hard. But it wasn't Enoch at all; it was her grizzled grandfather, Al. She was a child again. Al, always in need of a shave, had a scratchy face. His old blue work shirt with missing buttons had the same stains and rancid smell of age.

Sally squinted at the clock but couldn't make out the time in the night light they now kept for Enoch. She pulled the clock closer: 3 A.M. As her heart slowed, the dream faded and the day's memories rushed back. "Oh my God, Dr. Rudick." It can't be real. She must be misremembering. Sally became nauseated. Now, she could not avoid remembering the visit to the office. As Enoch stared into space, Dr. Rudick talked calmly about the need to make plans. He referred Sally to a social worker who could advise and help with arrangements for nurses and aides to visit the Wallaces and help with chores of daily living. He talked about "transitions" and "altered styles of life."

Sally was painfully awake now. Enoch was breathing regularly, without snoring. She had willed herself to listen carefully to Dr. Rudick, but his voice was far away. The office surroundings seemed thoroughly alien and unrecognizable. She froze in bed as the doctor's phrase suddenly reentered her consciousness: "We cannot exclude Alzheimer's disease." "Cannot exclude, cannot exclude, cannot exclude . . . " ran through her waking mind again and again. What a strange, impersonal way to talk about Enoch. Does "cannot exclude" mean "maybe," or "probably not," or "positively," or "positively not?" The doctors speak an incomprehensible language. Their words hide truth, or ignorance.

Enoch can't have Alzheimer's disease: just the other day they were discussing the election and Enoch led the way with his always insightful comments. But his forgetfulness, his absent-mindedness, his dishevelment were undeniable. Sally was gripped by a confused panic. Where could she turn? Who could help? Maybe the children would have ideas. Could she speak to Charlie? How could she speak to Enoch? What should she say? What is there to say when the neurologist, a world expert, says that he can't exclude Alzheimer's disease?

Dr. Rudick was as vague about timing as he was about excluding diagnoses. Enoch could be expected to pursue a "gradual course." How fast is gradual? How slow is gradual? What does a "course" look like? But as Sally thought back over the past two or three years, she recalled the early absent-mindedness, the rare periods of tangential, rambling conversations. Sally recalled Charlie's remarks about Enoch's forgetfulness at work, his episodes of disorganization. Then she thought about Enoch's legendary taste in clothes, the way he matched the novel tie with the unexpected belt and shirt: somehow it always worked in an understated, pleasing way. What a contrast with Enoch's recent sloppiness, his disregard for wrinkled shirts, clothing mismatches, undone buttons. "Gradual" might be a long time. Gradual.

Their early days in New York had been so new and fresh and exciting. They had always worked together. Sally began as an editor for the museum's magazine. Enoch reviewed the articles with her, never failing to offer the trenchant suggestion, the felicitous phrase, the insightful observation about a piece's structure. Sally, in turn, counseled Enoch on managing his new staff at the firm, handling Alex (the paranoid, jealous, junior associate), judiciously interacting with Gavin (the imperious, senior managing partner). Their interests, talents, and thought patterns were so complementary. Sally's love of language, her flare for expression matched Enoch's precise, logical, yet esthetic, mode of communication.

Sally loved the colorful, diverse group of editors at the magazine; she thrilled at the exotic New Yorkers rushing along Central Park West, each in his own world, studiously unmindful of the crowded street. This was definitely not the rural Connecticut of her childhood. The photo editor at the magazine, Mina, was a short, effusive, cultured packet of energy who was born in Iran. She spent time in every capital in the Mideast and spoke Farsi, Turkish, a smattering of Hindi, and a string of Romance languages. She had a new anecdote for every occasion, which she delivered in a mellifluous polyphony that thrilled Sally and Enoch. Though not a native New Yorker, she adopted Sally and Enoch, and introduced them to the nooks, crannies, kooks, and characters of the city. She knew every shopkeeper, restaurant owner, panhandler, and poodle on the Upper West Side. Mina transformed Sally's rural background and sheltered women's college experience into the quintessential American dream. For Mina, Sally, and Enoch, the mundane and miraculous melded.

As Sally sat up in bed, staring at the sleeping form of Enoch, she felt that she had let that wonderful life slip away. What happened? What had she failed to do? Through lack of vigilance, not living mindfully enough, not doing something, she was responsible for the present tragedy. But she wasn't even sure that there was a tragedy. It was all so confused. In bed, in the lonely, night-light dusk, Sally lurched from despair to hope to numbness. The rush of thoughts was chaotic. She couldn't concentrate. Remembering Enoch's acute observations today about interest rates and the economy, Sally brightened. But recalling his distant gaze at dinner, and his carelessness with the ice cream dessert, she froze. Would they need a nurse visiting daily, invading their home and imposing an antiseptic routine? Sally had never adapted to housekeepers, childcare workers, and baby-sitters. The building doormen still intimidated her. She hated strangers in her house.

The approaching loss of control was frightening. She couldn't face Dr. Rudick's "professional healthcare workers," which he described so reassuringly. How could he fail to understand? They were the markers of life's dissolution; they would be the architects of the loss of independence. The half-remembered, half-created image of the huge, leviathan nurse in white carrying her screaming to the operating room flooded her consciousness. As a four-year-old, she had broken her arm falling from the crab apple tree and was handed over to the nurse in the hospital by her trembling mother. She remembered the cage of a nose cone that delivered the sharp, penetrating ether. She felt the claustrophobic fear yet again. For fifty years, half a century, the fear that day in the hospital,—the monstrous nurse, the ether, the abandonment,—welled up at times of crisis and stress. Enoch's car accident, Roger's childhood febrile convulsions, her mother's death, her own biopsy, all mysteriously called forth the hellish horror of the hospital, nurse, and operating room. And now Enoch's debility, his disorder, his incapacity, his diagnosis that could not be excluded, awakened the demon again.

"Is it time to go to the beach?"

"Enoch, we're in Manhattan."

"Oh. What time is it?"

"It's 3:30."

"Well, it's not time to get up then."

"No, go back to sleep."

"Why are you up?"

"I was just thinking."

"When are we going to the beach?"

"We're going to the Vineyard this weekend."

Enoch turned over in bed and was breathing the easy cadence of sleep within seconds. Sally was alone again. The "gradual course" of Enoch's Alzheimer's disease involved both growth and regression, to be sure, but the forces of death were now far outweighing those of birth and creative development.

❖

Our protective preoccupation with the mundane in the West, and with Maya in the East, obscures the central truth that death is a constant companion who can visit anytime, anywhere. The ancients understood that death has many forms, and appears in many costumes. "I come as time, the waster of the people,"[1] or I visit as "The sickness unto death,"[2] or a reminder that "you and I Arjuna, have lived many lives. I remember them all. You do not remember,"[3] " I am the birthless, deathless Lord of all that breathes."[4] Biology has caught up. Biology is teaching that different deaths are, indeed, different.

Charlene Washington's stroke, described in Chapters 1 and 5, resulted from the unhappy association of two entirely different types of death. The sudden onset of right-sided weakness and slurred speech that hit her while she was conversing with Gary in her office, was caused by a blood clot lodging in a critical brain artery. The reduced flow of blood to the speech area and the motor areas on the opposite, left side of her brain deprived neurons of blood, oxygen, and glucose, resulting in instantaneous malfunction, and death of some cells within minutes. This initial, acute death largely resulted from oxygen deprivation (hypoxia). But Charlene's evolving deficits were also attributable to a second type of death, a delayed death.

The clot did not completely block all blood flow. A trickle variably flowed around the clot, allowing most of the neurons to hang on to life in a damaged state. But function was compromised, and the symptoms of weakness and speech impairment waxed and waned with blood flow over the next several hours. The wounded, dysfunctional cells participated in the secondary, delayed death. Hypoxia placed the neurons in an excitable state, causing them to fire wildly and release their neurotransmitter signals. Many of the damaged neurons released abnormal amounts of glutamate (glu), an excitatory signal that caused other neurons in the area to fire wildly in a furious neuronal chain reaction. In fact, toxic quantities of glu flooded out of the damaged neurons, leading to unbridled, raging excitation of neighboring neurons. The neurons were literally excited to death in this second round of delayed death. Acute death was due to hypoxia. The delayed death was caused by glu excitotoxicity.

❖

We now know that delayed, excitotoxic death is responsible for many of the devastating deficits caused by stroke. This realization is leading to entirely new treatments designed to prevent excitotoxic death, a revolution in stroke therapy. In turn, these insights have led to the discovery that delayed excitotoxic death is a feature of diverse disorders, including Alzheimer's, Lou Gehrig's disease, and brain and spinal trauma. How does secondary death actually occur? How does excitotoxicity work? What is the relationship of excitotoxicity to environmental toxins, to the food we eat? What is the role of secondary death in the genesis of Alzheimer's disease? How do we exploit our new knowledge to develop new treatments? These are some of the questions we now address.

Growth mechanisms are critical for synaptic strengthening, learning, and memory. The transmitter signal, glutamate, and its special coincidence detector, the NMDA receptor (see Chapter 8), are central molecular players. Glu and the NMDA receptor govern the development of synapses and circuits in the newborn and strengthen synapses during maturity. A collective sense of fulfillment accompanied the identification of some of the molecular architects of brain development and of memory mechanisms. The emerging commonality of developmental and memory mechanisms was an unexpected, unifying bonus. The optimism of discovery and insight was heightened by the subject itself. After all, this is about the happy circumstance of birth, growth, and the memories that enrich our lives and culture.

Glutamate and Excitotoxicity

A parallel scientific countertale was unfolding, however. This tale was also about glu, but the emerging, dark message was one of debility, death, and destruction. Glu, the creative element of growth, glu the agent of memory, also comes as the "waster of the people" that ripens to their ruin. Glu in excess, the uncontrolled actions of glu, turned out to be central to stroke and paralysis, epilepsy, the devastation of brain trauma, and even the dementia of degenerative disease. How is the agent of creative development and mentation related to brain destruction? How did that story evolve?

It has been known for decades that glu excites neurons and makes them fire. Glu is now seen as the main excitatory neurotransmitter in the brain. Approximately 80 percent of brain synapses that excite neurons to fire use glu as the transmitter signal. It is no surprise that glu inhabits the brain in high concentrations. For all these reasons glu has snagged the attention of researchers for decades.

The early experiments were puzzling, and perhaps even unbelievable. Forty years ago, infant mice were treated with glu to understand its actions. But, strangely enough, glu treatment *killed* neurons in the retina! The experiments were repeated and extended. Glu treatment also killed neurons in many brain regions. The scientific community was incredulous. The doubts, however, were met by a flurry of experiments by John Olney, a pioneer in the area. He demonstrated that oral or subcutaneous treatment with glu killed brain neurons in many species, including primates.[5] This was getting close to home.

To understand how glu exerted these deadly effects, Olney tested other chemicals that shared glu's properties. He found only chemicals that, like glu, excited neurons killed them. Chemicals sharing all of glu's traits except its ability to excite neurons did not kill. The association between excitation and death was so strong that Olney coined the term "excitotoxicity." Somehow glu and related chemicals excited neurons to death.

To confirm the excitotoxicity hypothesis, another classic strategy was employed. If glu and similar agents kill neurons through excitation, then chemical antagonists that block glu's excitation should protect the neurons. The antagonists prevented neuron death in the brains of mice injected with glu. Blockade of glu excitation also prevented the death of neurons in culture, indicating that the antagonists were protecting neurons directly; protective effects in the live animal were not due to indirect actions, such as altered glu metabolism. If specific chemicals protected against neuron death, new treatments for brain diseases could be at hand. The excitement about excitotoxicity was palpable.

All experiments seemed to support the strange, counterintuitive excitotoxic hypothesis. From folk wisdom to modern science, however, the whole idea of excitotoxicity had an implausible ring. Stimulate a muscle with exercise and it grows stronger, bigger, faster. Couch potatoes watch their muscles waste away through lack of exciting activity. Hence the lesson, "use it or lose it." An unbroken, honored lineage, from classical Greece to the moderns, of teachers, athletic coaches, and grandmothers, preached the truth that practice makes perfect. A cursory review of databases from classical Greece to the Latin world to medieval Europe to the Renaissance finds scant reference to the virtues of disuse, rest, and slothfulness. Yet the counsel for moderation in all things, for temperance, for the golden mean, has been another beacon for the fair species.

Excitotoxicity in the Environment

Glu, in the form of monosodium glutamate, or MSG, is one of the most widely used food additives in the world. Food regulatory agencies impose absolutely no restrictions on the addition of MSG to food; yet there is no doubt that MSG kills brain neurons. The developing brain of young animals is particularly vulnerable. Oral treatment of experimental animals with food doses of MSG kills specific brain neurons.[6] The affected neurons lie in areas unprotected by the blood-brain barrier, which excludes outside toxins. Hence MSG destroys a group of neurons in the hypothalamus, neurons that control hormone production and secretion throughout the body. After treatment of young animals with MSG, an endocrine deficiency disease gradually develops. We still don't know, though, whether a child who ingests MSG develops endocrine disturbances and disease later in life.

Many chemicals in the environment are closely related to glu and are also excitotoxic.[6] L-cysteine is, like glu, an amino acid and kills neurons in the immature rat brain. The pattern of brain damage closely resembles the pattern we see with decreased blood flow and reduced oxygen availability. Nerve cells die in the cerebral cortex and the hippocampus, resembling the destruction in strokes. Again, the developing brain is especially vulnerable. The stroke-like damage is nearly identical to that of cerebral palsy. Yet the relation of excitotoxicity to cerebral palsy, a deforming, life-long affliction, remains to be explored.

In 1987, a shocking incident of food poisoning struck more than one hundred people in Canada.[7] Many died. At autopsy, it was found that neurons throughout the brain were destroyed. Many of the survivors suffered from severe cognitive deficits. The hallmark deficit was memory loss. The patients had severe anterograde amnesia similar to H.M.'s (Chapter 8). They could not lay down new memories. All affected individuals had eaten mussels from Prince Edward Island that contained high levels of the excitotoxin, domoic acid. This chemical, also closely related to glu, excites neurons by binding to glu receptors. This is an instance in which an excitotoxin causes cognitive deficits not unlike those in the Alzheimer's disease ravaging Enoch Wallace. To add to the interest, a striking number of the severely afflicted individuals were elderly, and thus their symptoms mimicked Alzheimer's even more. All in all, the outbreak raised puzzling questions about excitotoxicity and cognitive deficits in brain disease.

Environmental excitotoxins can also induce strange degenerative neurologic diseases. Some residents of a group of South Sea Islands, particularly Guam, have developed a uniquely devastating degenerative disorder.[8] The unfortunate patients who contract this disorder have ingested the seeds of the cycad plant. The seeds contain the excitotoxin known as BMAA, another amino acid related to glu. The BMAA appears to exert destructive effects on neurons by interacting with glu receptors. This disorder brings us close to home and close to the tragedy of Enoch Wallace. Its gradual progression, widespread death of neurons, and degenerative nature are quite similar to the degenerative scourges of Lou Gehrig's, Parkinson's, and Alzheimer's itself. Could excitotoxicity play a role in Alzheimer's disease?

Tenuous hints derive from the crumbling tenements, alleys, mean streets, dank gyms, and broken-down rings of the boxing world. Walker Smith Jr., born into the poverty of Harlem, became known to the world as Sugar Ray Robinson, arguably the greatest fighter pound-for-pound in history. Robinson possessed a lethal grace, blinding speed, and a knockout punch with either hand. Yet his unparalleled boxing skills also allowed this willowy but muscular matador to dominate opponents with skill and strategy alone. He held world championships in two different weight classes in his career, and destroyed many ring immortals including Jake LaMotta, the Bronx Bull, Rocky Graziano, a deadly puncher from the Lower East Side, and Kid Gavilan, the Cuban Hawk. Sugar Ray became synonymous with physical perfection, combining near-impossible grace, speed, and power. When he retired from the ring, he leapt onto the stage and launched a career as a dancer and singer, employing the same dazzling grace. He died a helpless, demented man, decimated by Alzheimer's disease.

Sugar Ray Robinson succumbed to *dementia pugilistica*, a disorder that strikes fighters who have absorbed too many blows. In a remarkably similar experimental setting, concussive brain injury in the rat allows excitotoxins to escape from cells in the hippocampus and reach dangerous levels. The presumption, though not yet proven, is that repeated blows to the head release excitotoxins in boxers that stimulate receptors and lead to the progressive death of neurons. It is particularly provocative that a microscopic hallmark of Alzheimer's disease, "neurofibrillary tangles" in damaged neurons, are prominent in *dementia pugilistica*. Extending the circumstantial evidence, exposure of cultured human neurons to glu (or the related excitatory amino acid, aspartate) leads to similar microscopic findings. A direct link between glu, excitotoxicity, and Alzheimer's remains to be established. The relation of these observations to the Parkinson's disease now victimizing Muhammad Ali is unknown.

These few examples of environmentally induced excitotoxicity barely hint at the glu time bomb that dwells within.

Excitotoxicity Within

Environmental excitotoxicity is a pale reflection of the death dealt by unregulated actions of the glu within our brains. Glu, the developer of synapses, glu, the agent of synaptic strengthening, glu, the transmitter of learning and memory, is also glu, the final destroyer in many baffling brain diseases. Glu has been implicated in such wildly diverse brain diseases that the links, if any, seem hopelessly obscure. By examining different disorders, though, we can identify common themes that elucidate the Jekyll–and–Hyde nature of this brain signal.

Glu-induced cell death has been invoked in stroke, epilepsy, brain damage due to low blood sugar (hypoglycemia), brain trauma, and Alzheimer's disease. These disorders differ radically in symptoms, time span, age of patients victimized, areas of brain afflicted, predisposing factors and underlying causes (mostly unknown). How in the world can the single agent, glu, be held accountable for all these evils? To find out, I will try here to describe each disorder and extract the common threads, if any, that shed light on the Janus-faced mystery of glu.

In many ways, stroke, a leading killer in postindustrial society, seemed least in need of a hidden culprit. Strokes occur when the brain suffers diminished blood flow with falling oxygen levels—and the cells die. What could be more straightforward? Lack of oxygen leads to suffocation and death. Case closed. Not quite. In 1984 surprising findings were made in cultured hippocampal neurons that are particularly vulnerable to low oxygen concentrations.[9] The cultured neurons were exposed to dangerously low levels of oxygen that ordinarily were lethal. But one group of cultures was pretreated with drugs that prevent the excitatory actions of glu. The glu antagonist prevented low oxygen from killing the neurons. At the same time live rats were subjected to experimentally induced strokes that normally kill their hippocampal neurons. Yet the injection of a drug that blocks the glu NMDA receptors on neurons protected them during the stroke: they did not die.

In live rats, the lowered blood flow to the brain that occurs in stroke prompts the release of glu (and a related excitatory amino acid, aspartate) by neurons. Glu floods the local area and devastates neurons in the vicinity. The provisional sequence of events is diminished blood flow, followed by reduced oxygen, followed by damage to neurons that makes them fire wildly. Damaged and discharging neurons release lethal levels of excitotoxins; hence the massive cell death associated with stroke.[10]

From this sequence come several tentative lessons. Neuronal damage and uncontrolled firing rapidly releases toxic levels of glu. In turn, the glu rapidly kills neurons in the area. Glu markedly exacerbates the already devastating effects of compromised blood flow and diminished oxygen. Indeed, protection by glu antagonists suggests that excitotoxicity may be the final common pathway of brain cell death in stroke. The future use of glu antagonists to treat stroke is one happy byproduct of these scientific insights.

Other brain disorders characterized by uncontrolled neuronal firing may cause widespread cell death. Repetitive epileptic seizures, like those suffered by H.M., are of this type. In experimental animals, repetitive convulsions result in neuron death that closely mimics the deaths observed with glu toxicity. The pattern of neuronal death resembles that for patients who succumb during epilepsy. Drugs that block glu action protect brain neurons from death due to seizures, paralleling the results for strokes.

Links of excitotoxicity to Alzheimer's disease itself are tenuous but provocative. While many brain neurons degenerate in Alzheimer's, one group is of particular interest. Death of the basal forebrain cholinergic neurons that contact the hippocampus is largely responsible for devastating memory deficits. In experimental animals excitotoxin treatment destroys these same neurons. The clear implication is that these neurons are vulnerable to endogenous excitotoxins in the brain. Animals treated with excitotoxins exhibit such characteristic memory deficits that they have become standard models of Alzheimer's. Adding to the interest, one year after treatment with the toxins, their brains exhibit the typical neurofibrillary tangle of Alzheimer's. (In this fast changing field, though, the findings are controversial.) I already mentioned that treatment of human, as well as rodent, neurons in culture with glu also leads to the microscopic "plaques and tangles," that are characteristic of Alzheimer's; this adds fuel to the excitotoxic fire. Furthermore, excitotoxins other than glu have been implicated. The related toxin cys is elevated in Alzheimer's patients' blood. While tantalizingly suggestive of an excitotoxic contribution to Alzheimer's, the evidence hardly constitutes a medical smoking gun. The search continues.

Of the many mysteries, a pivotal one demands explanation. How can the single problem of excitotoxicity account for disorders as diverse as acute strokes, cataclysmic seizure damage, and slowly progressive Alzheimer's? This question can be approached only by understanding the nature of excitotoxicity. Why does the glu of brain growth and synaptic communication also kill?

How Glu Kills

Without glu, our brain neurons could not communicate; the brain would resemble a computer without power, a television without transmission, a silent pinball machine. But with too much glu in the wrong places the brain dies. Glu appears to kill via two related death pathways that represent life-giving processes run amok.

The first is calcium, the builder of bones and teeth, and an essential intracellular signal. Without calcium synchronizing signals in a web of cellular operations, cells die. Above a certain concentration, calcium kills.

The second pathway, of all things, is life-giving oxygen. Deprived of oxygen in a cardiac arrest, for example, the victim's brain has four minutes to live. After 240 seconds, irreparable damage leads to coma and a vegetative state. No amount of medical heroics after that point can restore brain function. However, as a recurrent refrain, too much oxygen or its related chemicals kill.

Glu can kill in a startling array of diseases, so understanding some common final pathways of cellular life and death is critical. These insights also provide the perspective to appreciate commonalities in Alzheimer's, stroke, and epilepsy.

Oxygen and Life: A Brief History

Even though oxygen is essential to life as we know it, this was not always so. Life originated on earth about 2.5 billion years ago in the absence of oxygen. Indeed, life could not have arisen in the presence of oxygen due to its toxicity. All of our cells carry the ancient heritage of vulnerability to oxygen toxicity, and glu can awaken the dormant destroyer.

As we learned early in our own lives, life itself consists of water, salts, and carbon compounds termed "organic" because they are almost exclusively produced by life. Taken as a group, water, salts, and carbon compounds comprise over 99 percent of all living material. The organic compounds of life fall into four familiar categories: carbohydrates, fats, proteins, and nucleic acids—constituents of genes. The nucleic acids, extremely complex molecules, are made up of strings of nucleotides in a nearly endless series of combinations that encode the heritability of life.

Specificity, or the selectivity of function and complexity, reaches astounding heights with the proteins. These are the largest and most complex molecules yet discovered. Their building blocks are simple. Proteins consist of about twenty-five different amino acids linked together in chains that are hundreds to thousands of amino acids long. The amino acid units are strung together in every conceivable order; chains branch from chains which branch from chains, virtually endlessly; whole huge molecules fold into biomorphic structures in three-dimensional space and form spheres, fibers, planes, helices, and tubes, replete with grooves, pockets, surfaces, hinges, and on and on in infinite variety. Adding another level of complexity, individual proteins aggregate to form multiunit structures, where higher and higher levels emerge. No two species contain the same configurations of proteins. Proteins are the structural building blocks of life, and are also the machines that physically build life.

As building blocks, proteins are critical components of the external cell membranes—the fibrous struts that support the membranes, the fibrous internal skeleton that governs cell shape and structure, the inner cell organs such as mitochondria, the cell's energy factories, the molecular motors that move cells and their constituents, the receptors that respond to signals, the signals themselves, and, last but by no means least, the nucleus that houses genes, which themselves form complexes with proteins.

Proteins are also the molecular machines that physically build life. Enzymes are the huge class of proteins that use water, salts, and organic compounds to construct life. In their exquisitely designed complexity, enzymes bring together molecules, form new associations between them, and create entirely new compounds for cellular use. Enzymes thereby synthesize the carbohydrates, fats, other proteins, and DNA necessary for life. Enzymes also digest nutrients, freeing energy and substituents on which life depends. This destructive process breaks down molecules to release energy for cellular use. Without enzymes, life stops—or at least slows profoundly. This is a crucial point.

How could organic compounds, and life itself, have arisen in the absence of enzymes? The answer is "time." Enzymes are catalysts that profoundly increase the rate of chemical reactions. Enzymes, for example, can increase (by a multibillionfold!) the rate of combining simple molecules into organic compounds. These same reactions also take place without enzymes. Yet a reaction that takes a millionth of a second when it is catalyzed by an enzyme may take seconds, hours, or days in its absence. The reactions may take eons without enzymes. The primordial planet lacked many facilities, but it most certainly had the required eons.

Over the course of the earth's eons, the chemicals of life were synthesized and life's components appeared. The absence of the twin destroyers, enzymes and oxygen, was critical. Hence new organic compounds were not immediately destroyed.

But how does life today continue in the presence of enzymes that cause decay? And how did oxygen the destroyer come to play such a central role in creation and destruction? Answers to these intimately related questions shed light on how glu kills consciousness through enzymes and oxygen.

The life-creating chemical reactions catalyzed by enzymes are all reversible. Enzymes increase the speed of synthesis of organic compounds in the forward direction, but they also increase the speed of decomposition in the backward direction.

It is a sad fact of life that decomposition is accomplished with much more ease than synthesis. The construction of a complex protein is a relatively slow, step-by-step, assembly, whereas destruction and decomposition occur with blinding rapidity. Somehow the two must be held in equilibrium for life to be possible. How can living organisms combat the lightning forces of destruction? How can life survive?

The answer is energy. With it, life tips the balance from decomposition and destruction toward synthesis and creation. Through the constant use of energy and material, living organisms maintain and renew themselves. The need for energy is absolute and constant. Organisms are specially designed for the constant use of energy. In a moment's absence of energy, the scales are tipped toward the forces of decay.

How does life get the energy to live? Primordial organisms, presumably arising in the oceans, derived sustenance from the seas' organic chemicals. In the absence of oxygen, the only chemical trick available for energy extraction was fermentation. In a familiar reaction, yeast ferments sugar to yield alcohol and energy. Animal cells also ferment sugar, but in this instance lactic acid and energy are the reaction's products. That energy constantly fuels the chemical reactions of life, balancing the forces of decomposition.

But fermentation consumed the nonrenewable resource of organic compounds that had accumulated over the eons of prebiotic planet earth. To fuel maintenance, growth, and reproduction, life had been living on time borrowed from the eons. With the destruction of the legacy of organic compounds, life was racing toward its dissolution. It was simply a matter of time before life based on fermentation had to end. Before that cold silence descended, life nearly reinvented itself.

As fermentation proceeded willy-nilly, carbon dioxide (CO_2), its waste product, flooded the oceans and atmosphere in extraordinary quantities. As life hurtled toward extinction, fermenting its organic heritage, a momentous threshold was crossed. Organisms evolved a novel capability that changed the fate of the planet. Life invented photosynthesis, allowing cells to produce their own organic molecules. Life was no longer enslaved to the prebiotic universe. Life itself had created a new world.

Using the energy of sunlight, organisms initially synthesized sugar from CO_2 and water. The same revolutionary process used ammonia and nitrates as sources of nitrogen to synthesize the immense variety of proteins, carbohydrates and fats, the array of organic compounds. Harnessing the energy of sunlight, living organisms were emancipated from dependence on preexistent organic chemicals. Life moved in a new and fateful direction.

Organisms could now produce their own physical substance and, through fermentation, could provide the sustaining energy. Fermentation is profligate, though. It is terribly wasteful, leaving most of the energy in organic compounds untapped. With fermentation, cells capture only a tiny portion of the energy potential of organic compounds. To compound the difficulties, fermentation produces highly toxic wastes, including alcohol, lactic acid, and formic acid. With enough time, the poisons of fermentation would turn the oceans, land, and air into lethal cauldrons. Earth would revert to lifelessness.

Hope laid in the fact that another byproduct of photosynthesis is oxygen. As O_2 spread through the oceans, an entirely new evolutionary niche became available. Life ultimately seized the opportunity and invented a revolutionary new method of capturing energy: respiration. Respiration is thirty times more efficient than fermentation in energy capture by cells. Respiration releases and uses all the energy available in organic compounds. Cells could access vast amounts of energy and avoid decomposition by using a minimum of their own organic substance. As an added bonus, the products of respiration, carbon dioxide, and water are innocuous.

With the advent of cellular respiration, life moved to center stage on the planet. Respiration freed life from a marginal subsistence level. Relying on fermentation, life forever faced extinction. Even photosynthesis could scarcely keep up with the insatiable demands of fermentation. At best, fermentation barely supported the holding operation of survival. In striking contrast, respiration brought about a surplus of energy and moved life from the edge of annihilation. The energy of respiration facilitated the grand evolution of life.

Glutamate Toxicity and Oxidative Stress

Respiration releases and captures energy by way of oxidation, the chemical reaction of oxygen with organic sources of energy. During oxidative metabolism, oxygen attacks organic molecules and breaks chemical bonds, thus releasing energy. Yet oxidation generates very toxic byproducts called "oxygen radicals." Highly reactive, they can attack an array of cellular constituents and cause damage and death. These radicals include the deadly molecules, superoxide anion and hydrogen peroxide. In aggregate, the destructive actions of the radicals are termed "oxidative stress." Glu can exert many of its damaging effects through these radicals. The brain is particularly vulnerable.

Unlike other organs in the body, the brain derives virtually all of its energy from oxidative metabolism. As it does so, it generates oxygen radicals. Vast amounts of energy are required to maintain neurons in a state of electrical readiness and to drive the firing of millions of neurons. Moreover, many enzymes in the brain, such as the ones which synthesize the neurotransmitters—dopamine, norepinephrine, and epinephrine—generate hydrogen peroxide as a byproduct. Peroxide produced by these and other brain enzymes decomposes into hydroxy radicals, which are toxic in the extreme. The brain, then, constantly generates oxygen radicals and is thereby relentlessly threatened by oxidative stress.

The oxygen radicals attack proteins, fats (lipids), and DNA. Recall that proteins comprise the structural building blocks of cells and the enzymes that govern cellular chemical reactions. Fats are critical components of all cell membranes, which are integral to cellular function. DNA, it goes without saying really, is life's hereditary blueprint.

A thin line of defenses protects cells against the oxidative onslaught. Several vitamins, chemicals, and enzymes reduce oxidant concentrations. Vitamin C (ascorbic acid), a water soluble vitamin located within cells, is derived from citrus fruits and decreases radicals. In the forbidding world of environmental toxins, oranges, lemons, and limes remain user-friendly natural resources. Vitamin E (alpha-tocopherol), localized to cell membranes, also reduces the level of oxidants. Vitamin E deficiency causes peripheral and central nerve abnormalities due to cell membrane breakdown. This may result in peripheral neuropathy in which numbness, tingling, and weakness of the hands and feet are particularly pronounced. Large doses of the vitamin protect against atherosclerosis in which oxidation of cholesterol contributes to the clogging of arteries.

A simple compound, glutathione, is synthesized in the cell and exerts powerful protective effects. Glutathione consists of three amino acids hooked together that participate in multiple reactions in cells. Its most important function may well be that of an oxygen radical "scavenger." The glutathione scavenger efficiently mops up oxygen and the deadly hydroxy radical.

A final group consists of various enzymes that protect cells from oxidative stress by numerous defense mechanisms. All reduce the levels of oxygen radicals. One family of enzymes, termed SOD (superoxide dismutase), has achieved notoriety because its deficiency has been invoked as contributory to Lou Gehrig's disease. The SODs decompose the superoxide radical. Two other cellular enzymes, catalase and glutathione peroxidase, exert allied actions; these combine to degrade the superoxide radical and hydrogen peroxide. Together these enzyme groups constitute a first line of defense. In sum, brain cells have evolved defenses against oxidative stress, but the threat remains ever present and can be unleashed by any circumstance that causes excessive glu release.

How does glu unleash free radical attack? We do not yet know for sure. Clues suggest that excess glu initiates a chain reaction that eventually forms oxygen radicals. The deadly pathway leads through calcium. Interaction of the glu key with its receptor locks enables the active, electrically charged calcium atom to enter cells. Carrying two positive charges, it is represented as "Ca^{2+}," the "calcium ion." In addition to triggering the entry of Ca^{2+} into cells from the outside, glu stimulates other receptors that release Ca^{2+} from intracellular stores. To sum up, excess glu increases cellular Ca^{2+} to dangerous levels in neurons through several mechanisms. The calcium ion, in turn, elevates oxygen radical concentrations.

This ion is an intracellular signal messenger in cells. It translates extracellular signals like neurotransmitters into changes in cellular function. Among its many intracellular activities, Ca^{2+} activates groups of enzymes that form oxygen radicals. Some enzymes break down proteins; others metabolize fats; still others manipulate small organic molecules. All produce oxygen radicals as byproducts. The oxygen radicals attack cell constituents; then, in a vicious chain reaction, damage leads to the uncontrolled release of additional glu. The whole cycle begins again, and the destruction spreads like wildfire. The feed-forward reaction can rapidly destroy entire populations of neurons in the vicinity. In a way, then, glu run wild reexposes the ancient vulnerability of cellular life to oxygen toxicity. The brain cannot escape the legacy of oxidative stress inherited from the origin of life.

Elevated Ca^{2+} also activates entirely different enzymes that begin digesting the cell's substance. The Ca^{2+} second messenger activates families upon families of enzymes. While all of these enzymes perform critical functions in normal cells, uncontrolled activation is devastating. Indiscriminate activation of enzymes that normally break down used proteins destroys normally functioning cellular proteins and deranges cellular structure and function. Enzymatic metabolism of fats begins destroying cellular membranes that compartmentalize cellular processes—whence cometh operational chaos. The calcium ion also activates enzymes that attack DNA and destroy genetic material. The blueprint for all cellular operations necessary for life disintegrates. Each of these derangements further damages the glu storehouse in cells and results in the leakage of more glu into the cellular environment. On the devastating cycle goes . . .

While I have concentrated on acute effects of glu toxicity that can transpire in minutes to days, events may also unfold more slowly. Current opinion has it that borderline levels of glu elevation, minimal increases in Ca^{2+}, minor increases in oxygen radicals, and slight rises in enzyme activity over time can elicit chronic disease. Alzheimer's, Parkinson's, Lou Gehrig's, and Huntington's diseases fall into this category. At present, though the evidence is suggestive, the possibilities remain speculative.

Alzheimer's and Chronic Degenerative Neurologic Disease

The slowly progressive yet deadly character of the neurodegenerative diseases is mystifying. Over the years, selected populations of neurons in the central nervous system degenerate and ultimately die. Symptoms vary according to which neuronal populations are affected, but the central feature of these diseases is unrelenting neuronal death that leads to dementia, debility, and death.

Could these progressive disorders result from excitotoxicity? Though it seems unlikely at first glance, odd pieces of an ill-fitting puzzle are suggestive. As is often the case, fragmentary hints derive from other diseases that share features with degenerative disorders. Two extremely rare disorders are associated with elevated levels of excitatory amino acids, and other disorders are attributable to the chronic ingestion of excitatory amino acids. Huntington's disease, a bona fide degenerative neurologic disorder, has been connected to excitotoxic amino acids. By briefly reviewing these disorders we can perceive the tantalizing but tenuous links of excitotoxicity and neurodegeneration.

Sulfite oxidase deficiency is a fatal childhood disorder that leads to degeneration of neurons throughout the nervous system.[11] It is associated with elevated levels of the excitatory amino acid, L-sulfo-cysteine.

OPCA, a form of "olivopontocerebellar atrophy," is a recessive genetic disease that destroys parts of the brainstem, the autonomic part of the nervous system that connects the brain and spinal cord.[12] It is accompanied by a deficiency in an enzyme that breaks down the excitatory amino acid glu. OPCA is an adult disorder that ultimately prompts the cerebellum to degenerate; this region of the brain is critical for coordination, walking and balance, and motor systems that govern muscle function. While these medical rarities are provocative, no one knows whether increased levels of excitatory amino acids cause the disorders—a fascinating but unproved hypothesis.

A causal link is being forged for two other diseases in which ingestion of excitotoxins induces degeneration. The first is one we have already described: the disorder, afflicting victims in the Guam and Mariana South Sea Islands, combines features of several degenerative disorders and is associated with the ingestion of the excitatory amino acid BMAA contained in the endemic cycad plant.

The second disorder also kills motor neurons that control muscle function; it is called lathyrism.[13] It is associated with the chick pea excitotoxin, BOAA.

The association of two identified environmental excitotoxins with neurodegeneration is certainly suggestive. Perhaps excitotoxicity does contribute to chronic, progressive degeneration of neurons. What is more, the association of excitotoxins with the neurodegenerative group of diseases may be a big clue. The close association of ALS, PD, and features of Alzheimer's implies that the diseases are somehow related, even if excitotoxicity itself is not the single cause. Mid–life degenerative neurologic diseases that seem different on the surface may share commonalities that uncover cause and treatment. By identifying the mechanisms of brain disease, we find new targets for treatment. How do these insights lead to treatments that may render the brain a sustainable, if not a renewable, resource in the twenty–first century? That is the subject of the next chapter.

❖

With the home visit of the social worker, Sally could no longer deny that Enoch required professional care. The sense of control, the privacy of family, and the hopes for Enoch's recovery were sad illusions. In fact, it had been recognized for millennia that the brain does not recover after illness and injury. The inability of nerves to regrow was a tragic central fact of brain science. Yet, miraculously, in the latter twentieth century, revolutionary discoveries in the United States, Sweden, and Canada indicated that brain nerves can regenerate after all. These dramatic insights led to the experimental transplantation of nerve cells to the brain, and a radical reformulation of possibilities for treatment of brain and mind.

❖

10

THE BRAIN AS A
RENEWABLE RESOURCE

❖

WEDNESDAY WAS the appointed time. The case man-
ager or social worker or nurse, or whatever person Dr. Rudick designated, was to visit for a conference, a chat, to get to know the Wallaces
and their needs. It made perfect sense, but Sally was terrified. Enoch was
not quite clear about the purpose of the meeting.

Ms. Leah Sokoloff, a warm, energetic, soft-spoken woman, immediately instilled a sense of relief in Sally. Enoch seemed interested
enough, without reacting effusively or indifferently. Leah sat on the love
seat, opposite Sally and Enoch, across the coffee table, opened a worn
spiral notebook and started an easy conversation about her trip crosstown, the approaching summer, and their attractively decorated home.
The conversation gradually shifted to Enoch's interests, activities, and
plans for the summer. Sally's anxieties temporarily lifted as Ms. Sokoloff
imperceptibly transformed into an inquiring advisor, a tour guide planning a vacation, a hostess of sorts. Yet Ms. Sokoloff needed some routine

information, and Sally floated from hopefulness to anxiety and back again, fragilely hanging on every word and trying to resist reading Ms. Sokoloff's every look.

"What do you enjoy most about your home, Mr. Wallace? Do you have a favorite room?"

"I suppose the den is best. The easy chair is my favorite. Sometimes in the office, I think of the den and chair. I think of sitting with a martini, reading the paper or watching the news."

As Enoch went on, conversing effortlessly, obviously engaged with Ms. Leah Sokoloff, Master of Social Work, Sally experienced a new feeling. Was it distance, envy, or just insecurity? Enoch as foreign? Was she actually jealous of a social worker she had never seen before? Enoch seemed to enjoy the attention. He went on about the den, his work, their Vineyard house, details of the apartment. Sally hadn't heard him so fluent and animated for weeks. Ms. Sokoloff listened, took notes, but didn't say much.

Sally experienced an undefinable happiness while listening to Enoch as host and not patient, but she envied his attentions to Ms. Sokoloff even though she appreciated the social worker's skilled questions. The simple emotions of the past had receded over a year or two as they gave way to fear, relief, anxiety, hope, sadness, and satisfaction—part of each moment of each day. Nothing was clear anymore. Sally could never decide whether it was a good time, a bad time, or an indifferent one.

"How do you manage around the house, Mr. Wallace? Do you need reminders to get things done? You said you like to cook. Do you cook alone, or does Mrs. Wallace join in?"

"I cook alone. Sometimes Sally helps. Sometimes I need help with ingredients. Usually, I can find what I need."

Hints of the ever present confusion were beginning to sneak into the conversation.

"What have you cooked most recently?"

"I made Steak Diane, or was it the coq au vin? I'm not really sure. Sally, which was it?"

"Enoch, you grilled those wonderful salmon fillets at the Vineyard."

"Oh yes, the salmon fillets."

"What was that like, Mr. Wallace?"

"I don't completely remember. There were some difficulties. But I don't remember the entire meal or evening."

Sally surprised herself by interjecting that Enoch forgot that the salmon had to be marinated. He then couldn't remember exactly how to mix a marinade. He couldn't remember the soy sauce, could not recall where the sauce was stored, forgot about the ginger, and was confused about how to mix the ingredients. He spilled the soy sauce on himself, and dropped the salmon on the lawn. He then burnt the whole preparation on the grill while Sally was in the house for a moment.

Sally was shocked by the nearly hysterical rush of words pouring from her own, disconnected mouth. She felt the tears in her eyes, the frustration, the helplessness.

Ms. Sokoloff calmly took notes, without any sense of judgment or surprise.

"Did the meal finally work out?"

"We salvaged a very good salmon dinner," Sally indicated.

Enoch could no longer recall any of the details. What started as a simple description of a Vineyard grill ended in a jumbled confusion for Enoch.

"Mr. Wallace, who did the shopping for the meal?"

"Sally and I went to the supermarket together."

Sally silently recalled the scene in vivid detail. Enoch seemed small and overwhelmed by the cavernous market, the endless jumble of sections, the labyrinth of aisles.

"Let's look at the salmon first, it will be fresh," suggested Sally, as she instinctively headed for the counter behind the vegetable section at the back of the store. Enoch nodded, but then tentatively wandered toward the delicatessen section.

"Enoch, this way."

"Oh, of course." But Enoch searched Sally's movement and body language for clues as to direction. He followed Sally too closely: a little boy hanging on to his mother in a crowded store, trying desperately to keep up, to avoid the dread of being lost.

At the counter, Sally took the number, waited in line, and examined the fish laid out on the chipped ice.

"Those steaks look good."

"Enoch, that's the swordfish. The salmon is over here."

"We could have the swordfish instead."

"Enoch, we planned the whole meal around the salmon. Charlie loves grilled salmon." Then, slightly raising her voice in frustration, "Don't you remember our conversation with Charlie at all?" She was flooded with guilt before the words were fully articulated. Unaccountably, she was on the verge of tears . . . yet again. No relief from the chain of frustration, anger, guilt, and isolation that were lying in wait for every idle thought, every errant phrase. Life had become a trap. Enoch stared at the dead fish on ice, unaware of Sally in her prison.

"Mr. Wallace, did you need help shopping? Were you able to find all the items alone?"

Sally burst into bitter tears and ran out of the room. Enoch looked utterly confused, feeling that a foreign scene was being enacted outside of his life. He looked to Ms. Sokoloff for guidance, for orientation, for some hint of the incomprehensible events swirling around him.

Ms. Sokoloff grasped the situation immediately.

Sally returned composed, with a sense that the worst of the interview was over. The social worker continued in easy banter, asking general questions about Martha's Vineyard, and gradually steered the conversation back to the shopping trip. Enoch seemed anxious to resume his description of the supermarket, much to Sally's surprise. Of all things, he focused on the checkout counter; that business transaction had been anything but easy.

Sally and Ms. Sokoloff remained silent as Enoch launched into an unsolicited narrative.

"We were standing on a long line at the cash register. It seemed to go on forever. There was a whole row of girlie magazines. I remember the girls. One looked like Enid. I couldn't believe it. These were sex magazines. You could see everything. What was Enid doing on a magazine cover?"

"Enoch, *what* in the world are you talking about?"

Sally was dumbfounded; she had not heard a word of this before. The connection of their daughter-in-law and girlie magazines was so absurd, so thoroughly unimaginable, that Sally again descended into her now characteristic terrified and confused state.

Just as suddenly, Enoch dropped the subject, and began talking about paying the bill. Sally was again disoriented. Ms. Sokoloff listened calmly and silently.

"So we emptied the buying wagon and piled the food on the thing."

"Enoch, the 'shopping cart,'" Sally interjected unconsciously.

"That's what I said. Something happened. I forget. Oh yes, she added the charges wrong. She made a mistake. So we talked to the charge girl and straightened it out, and had to pay the ticket."

Sally was desperate. Enoch was living in another world. His story was sheer fiction.

In reality, when Enoch was given the register receipt, he presented the sales clerk with a jumble of crumpled single dollar bills, change, a credit card, and some papers on which notes were scribbled. Sally quickly intervened, paid the bill, and they beat a hasty retreat. Once outside, Sally could breathe again.

After a brief period of silence, Ms. Sokoloff began softly.

"It sounds like you have a wonderful life in Martha's Vineyard. Now we should think of what would be most helpful for you to have that kind of life here in the city."

"How do you manage in your city home, Mr. Wallace? Can you find all the rooms? Do you ever get lost? Can you find everything you need in the bathroom? Sometimes it's hard getting into the bathtub. It can be slippery. Have you ever fallen?"

In her embarrassment and anxiety, Sally marveled at Ms. Sokoloff's competence and tact. Without being explicit, the social worker touched on multiple areas of difficulty and shame. Somehow, they were now talking about bathrooms, personal hygiene, bowels and urination, without whispering, without using code words, without denying the years of humiliation. Sally experienced a sense of relief so foreign that she almost began crying again.

"Mr. Wallace, are you able to use the bathroom alone? Or is it easier to have someone there to help?"

Against her will, Sally held her breath. In spite of herself, she feared Enoch's response. Scenes of Enoch urinating in bed were more frequent now. He always seemed to be unaware. After coaxing and argument with an oblivious Enoch, Sally finally convinced him to place a small diaper in his underpants. In the bathroom, Enoch had urinated on a toilet bowl with the top down on occasion. Sally was never completely sure that Enoch was aware. Finally, they could talk together with someone else about it. Sally was grateful, and Enoch didn't seem bothered in the least for the moment. With a few deft questions, Ms. Sokoloff determined that there was no bowel problem. He moved his bowels every morning; he experienced no sense of urgency; his underpants and bed sheets were never soiled. She also noted the minor personal care and hygiene issues that arose.

"On the whole you seem to do very well around the house. Mrs. Wallace, maybe you could use a little help. Would a visitor several times a week for a few hours be helpful?" Ms. Sokoloff decided that limited home care was the only need at the moment, and that Enoch did not require placement in an adult care facility. He was surrounded by a caring family, obviously had the means to afford his present requirements, and needed only partial help. He did not require constant companionship, supervision, and was in no danger of harming himself.

❖

After seven years of L–DOPA–purchased freedom from the Parkinson's prison, the bars began closing in on Jeanette McCready (Chapter 3). A number of old symptoms reappeared, and new difficulties arose. The arm and leg muscle stiffness, which had been controlled but not eliminated by treatment, grew noticeably worse. The stiffness seemed to slow movement, which now required a supreme effort. In fact, all movement was slowed, and Jeanette felt as if she was moving through maple syrup. She had increasing difficulty arising from a chair, which now consumed seconds, and often resulted in a stumbling loss of balance. It was a fight just to stand up. The coarse hand tremor remained unchanged: the L–DOPA had never helped it anyway.

But the strangest, most frightening experience was what her doctor called the "on-off phenomenon." Though Jeanette could get around, with difficulty, most of the day, she had attacks of near paralysis that could last from minutes to half an hour. The spells came on suddenly, froze her in her chair, prevented almost all movement, and then mysteriously dissolved. The spells had something to do with the failing effectiveness of L–DOPA, but it was a medical mystery. Jeanette felt herself turning into a permanent cripple in spite of the early wonders of L–DOPA therapy. The miracle treatment was failing.

While L–DOPA therapy replaces the dopamine signals that decrease due to the death of "Parkinson's" brain neurons, treatment does not prevent the underlying progressive, slow, death process. L–DOPA provides signal for the surviving neurons. In time, however, as more neurons die, signal replacement for the few, remaining, debilitated neurons is simply ineffective. The underlying disease overwhelms treatment. Jeanette McCready was now a prisoner beyond the help of L–DOPA.

Jeanette had read in *The Boston Globe* about new medical research that was being performed on mice and rats with experimental Parkinson's disease: something to do with "brain transplantation." It sounded like Frankenstein fantasy to Jeanette, and she was afraid to bring it up with Dr.

Cottone. He was like a god, and she didn't want to sound desperate or as if she was going crazy or had become a weirdo. Nevertheless, after anxious anticipation and rehearsal, she broached the subject. He assured her that such experiments were not crazy and that there had been remarkable success in the laboratory, although it wasn't quite "brain transplantation."

Neuroscientists in several laboratories across the world were injecting neurons into the brains of mice and rats. They were using the neurons to replace those that had died in experimental Parkinson's disease. Miraculously, replacement with foreign, donor neurons largely cured the mice and rats. The "transplants" relieved the muscle rigidity, difficulties walking, and slowed movement. Though far in the future, the dream, of course, was to apply this grafting procedure at the bedside to patients like Jeanette, with end-stage Parkinson's disease.

How were the transplanted neurons curing the disease? Were they actually growing in the recipient brain? Were they forming new circuits? Or, were they triggering some other, unknown processes in the brain? Could this strange approach actually be used in people? Could it be used in Jeanette?

❖

Neurology—indeed, all of brain science in its full complexity—has helped to identify a few eternal truths. These immutable laws were often pessimistic and not terribly illuminating. One article of faith simply held that nerves do not regrow. Unlike virtually every other tissue in the body, nerves do not regenerate and recover function after illness or injury. This observation, documented over millennia, seemed to be one of the few statements about brain function that could be made with confidence. This natural "law," however, offered little or no insight into brain function.

The mystery of regrowth failure was periodically revisited and pondered as neuroscience emerged over the last century. Nearly a century ago, the great neuroscientist Ramón y Cajal alluded to rare instances where injury to the spinal cord or brain was followed by growth of new nerve processes in damaged areas. Such statements were regarded as little more than freakish exceptions against a background of overwhelming evidence that nerves do not regrow. Even so, the problem of (non-) regrowth periodically attracted attention, if only for violating all the rest of cell biology.

Half a century ago, in the 1940s and 1950s, the pioneering neuroscientist, Mac Edds, of Brown University, reexamined the problem with intriguing but mixed results. He was encouraged to reinvestigate the ancient, "hopeless" issue by the advent of new approaches, and by the advice of a colleague, Hughes, that, "since . . . cell biologists are on old ground working with new implements it would be well if they recognized more fully the antiquity of their sites and looked out for the old forgotten tracks, along which something of value might still be found."[1]

Working in the accessible peripheral nervous system, Edds cut nerves innervating patches of skin or muscle.[2] He found that some cut fibers could reinnervate the skin and muscle; even adjacent, undamaged fibers could sprout new processes that innervated skin and muscle deprived of nerves. Apparently, then, some nerves could partially regrow, at least in the peripheral nervous system. Whether this was remotely relevant to nerves in the central nervous system, the brain and spinal cord, was doubtful.

Stimulated by Edds's results with peripheral nerves, neuroscientists Chan-Nao Liu and W. W. Chambers at the University of Pennsylvania Medical School in Philadelphia decided to reexamine the possibility of regeneration in the central nervous system.[3] They chose the spinal cord because it presented a unique combination of scientific advantages. As part of the central nervous system, any discoveries might be relevant to the brain, yet the spinal cord was more accessible and less complex than the brain itself. Also, peripheral nerves from all over the body connected to the cord, conveying sensation. So, the cord offered a unique opportunity to study peripheral nerves and the central nervous system at the same time. The symmetric architecture of the sensory nerves and cord also simplified things. The cord runs from head to tail like the trunk of a tree. Sensory nerves in the body enter the cord from the right or left sides, like branches on a perfectly symmetric, two-dimensional cartoon Christmas tree. Because of this orderly arrangement, carefully controlled experiments can compare left nerves with their matched, right-sided partners, or compare adjacent nerves on the same side. Having recognized the advantages, how would one test for the possibility of regrowth?

Liu and Chambers decided to cut one of the peripheral sensory nerve bundles in the body just before it enters the cord and search for evidence of regrowth.[3] This straightforward approach had several advantages. By cutting the nerves outside of the cord, the experimenters avoided damaging the central nervous system itself. Earlier studies had already defined the precise nerves inside the cord to which the entering sensory nerves connected. Thus Liu and Chambers knew exactly where to look for the possibility of regrowth and reconnection.

Under sterile conditions and anesthesia in the operating room, the nerves of cats were cut. (They received optimal postoperative care and recovered from the minor surgery.) The investigators examined their spinal cords at various times after surgery. As expected, the surgery led to degeneration of the nerves beyond the cut, as they entered the spinal cord. The degeneration caused denervation of the intraspinal neurons to which the entering fibers normally connected. No surprises yet.

Liu and Chambers examined the spinal cords for regrowth after surgery. The average postoperative period was approximately three hundred days. In cat after cat, the results were dramatic. Examining the cords microscopically, the researchers found growth in every instance. The new growth did not come from the cut nerve, however. Rather, the nerves adjacent to those that were cut exhibited abundant sprouting. The fibers of the adjacent intact nerves sprouted new processes like small branches from the main branches on a tree. The degree of branching depended on how much degeneration there was after the cut. The more the damage, the greater the branching of adjacent fibers. Somehow, cutting nerves and degeneration stimulated renewed growth of adjacent peripheral nerve processes within the central nervous system. No distant nerves sprouted.

All right, then: nerves can regrow within the central nervous system. Although this was an important discovery, the fact remained that the sprouting nerves themselves were peripheral, not central. What was the next step? If degeneration of neighboring peripheral nerves induced sprouting, what would be the effect of cutting central nerves within the spinal cord? Liu and Chambers could ask this question because of the anatomy of the brain and spinal cord. In addition to the sensory nerves entering the spinal cord from the body, the brain itself also sends long nerve fibers into the spinal cord. One large bundle of nerves on either side of the cord descends from brain to cord and innervates neurons in the same areas as the incoming sensory nerves just described. These are the motor neurons that govern voluntary movement throughout the body. Employing delicate neurosurgery, Liu and Chambers cut these descending fibers and induced degeneration. The degeneration denervated the same intraspinal neurons as before. The effects were striking. Again, the neighboring sensory neurons entering the cord from the body sprouted vigorously. Again, only those sensory nerves adjacent to the damaged motor fibers sprouted. These experiments indicated that damage to peripheral or central nerves—and the subsequent denervation of spinal cord target neurons—can elicit new growth of neighboring peripheral neurons within the central nervous system. A critical obstacle had been overcome. The general phenomenon was named "collateral sprouting" to denote the regrowth of nerves in the vicinity of damaged neurons. The potential implications were at least as exciting as the dramatic results themselves. The experiments strongly hinted that something, perhaps some chemical signal, in the environment of degeneration triggers nerves in the spinal cord, the central nervous system, to regrow.

As Liu and Chambers performed their pioneering experiments, a team in Gainesville, Florida was also attracted by the discoveries of Edds. James Horel, a noted neuroanatomist, and his associate, Donald Goodman, at the University of Florida, employed an ingenious, complementary approach.[4] They worked in the brain itself, taking advantage of features of the visual system. Neurons in the retina sense an image and convey impulses to a relay station in the brain. In turn, the relay station neurons connect to the visual cortex at the back of the brain. In a feedback arrangement, the visual cortex neurons send fibers back to the relay station where the retinal fibers connect. Once again, then, a group of neurons in the central nervous system is a target for two neural inputs. Goodman and Horel wanted to know whether destruction of one of the inputs to the relay station caused the other input to sprout. They destroyed the visual cortex on one side of a rat brain, which deprived the relay station on that side of the input from these neurons. After 16 months, they examined the brain relay stations. Vigorous sprouting of fibers from the retina occurred only in relay stations deprived of visual cortex input.[4] The relay station on the opposite side, that was contacted by the normal complement of control, uncut fibers, exhibited no sprouting, as expected. Goodman and Horel had demonstrated that neurons in the mammalian brain could regrow, in agreement with Liu and Chambers's results.

So here we have the partial story of regrowth: Destruction of one group of neurons somehow stimulates the regrowth of other nerve fibers in the area. Regrowth occurs in the brain and spinal cord. Regrowth occurs in rats and cats. The potential for regrowth was beginning to look like a generalized capacity. Sprouting was transforming from a freakish sideshow to a real phenomenon.

As work progressed in the States, a young neuroanatomist at Oxford University was taking a quantum leap. Geoffrey Raisman was the quintessential eccentric British scientific genius.[5] Raisman used the electron microscope to ratchet the study of regrowth down to the level of individual synapses. And he applied quantitative methods that permitted conclusions well beyond those available to his colleagues in the United States. Raisman is a full-time scientist, part-time novelist, expert in oriental philosophy, and gifted calligrapher. He often graced his friends and collaborators with calligraphy that synthesized the exquisite oriental line with a transporting depiction of brain anatomy. In one inspired work, he combined original drawings of neurons by Ramón y Cajal with his own calligraphic art that nearly created a new form, neurocalligraphy.

When he left Oxford to set up a new laboratory at the famed Mill Hill scientific center in London, he attracted an unusual collection of talents. Not the least unusual was Nigel. Visitors to the laboratory were invariably impressed by Nigel, who frequently accompanied Raisman on his scientific trips, and who often tagged along when scientific visitors were taken to dinner by the Raisman lab. Nigel, at the time, was a 4-foot boa constrictor, an integral member of the laboratory. Nigel had free run of the laboratory and frequently accompanied Raisman guests and visitors to the local pubs to discuss the scientific problem at hand. Nigel was not always immediately introduced to a visitor since there was nothing particularly noteworthy about his presence in a nondescript satchel in a car's back seat. More than one visiting scientist expressed mild surprise, moderate anxiety, or sheer terror when Nigel was properly introduced at the pub. Their consternation never failed to amuse the laboratory's pub-hopping members.

Raisman performed his experiments wholly on the brain as he searched for the potential for regrowth of tracts that were completely central. The experiments thus differed from those on the spinal cord, which concentrated on peripheral sensory neurons, and even from those of Donald Goodman and James Horel, which examined nerves from the eye that technically lie outside the brain. Raisman chose a group of neurons in the septum, deep in the base of the brain. The septum sends nerve fibers to the hippocampus, one of the key memory centers. (Though not recognized at the time, the septum and hippocampus degenerate in Alzheimer's disease; see Chapters 3 and 8.) As in the preceding approaches, the septum is innervated by two separate and distinct neural inputs, so nerve interactions can be examined in detail.

Using the electron microscope, Raisman wanted to know whether individual synaptic connections among neurons changed when one of the pathways innervating the septum was destroyed. He destroyed one of the neural tracts innervating the septum and examined the effects on synaptic connections. He scrutinized brains from one day to one year after surgery. After more than 100 experiments were carried out and more than 50,000 individual synapses were classified, counted, and quantitated, the results were mind boggling. The destruction of one pathway produced "vacated synapses." That is, the presynaptic terminals disappeared, but the postsynaptic part of the connection was recognizable as the "vacated synapse." Over time, the vacated postsynaptic structures became innervated by nerve fibers from the remaining, undamaged pathway. Sprouting and regrowth thus resulted in synaptic reconnection. Sprouting and regrowth were leading to reconnection and true regeneration. By quantitating the results, Raisman demonstrated that the vast majority of vacated synapses became reinnervated.

Raisman concluded that something about the vacated synapses themselves stimulated renewed growth in the remaining pathway, and regrowth of the synaptic connections. He was drawing one step closer to the secret of the regulation of growth and synaptic connections in the brain. Moreover, this was a neural pathway in the brain that played a central role in learning and memory.

In a combination of British understatement and barely contained optimism, Raisman concluded that "the central nervous system can no longer be considered incapable of reconstruction in the face of damage. . . . Central synapses are labile, and this plasticity may not be restricted to lesion [*damage*] situations—it may play some part in learning."[5]

In the last sentence of his scientific article, he speculated on the lack of regeneration in the brain, and implied strategies for therapeutic approaches far in the future: "The rapid re-occupation of sites by . . . local terminals may remove an important stimulus for regrowth of the original cut axons [nerves] and thus be a factor in the apparent failure of effective anatomical regeneration in the central nervous system."[5]

Raisman suspected that the remaining, vacated synapses were releasing substances that attracted neighboring nerve fibers. Once the new connections formed, the new fibers somehow used the substances and removed any triggers for growth of the cut nerves. Though not mentioned explicitly, in a prescient leap, he was anticipating the discovery of growth and survival factors in the brain, and even dreaming of new treatment strategies.

What had begun as infrequent, Quixotic forays was transforming into a serious pursuit. The possibility of central regrowth, though still remote, carried such profound implications that an international conference was held. The National Paraplegia Foundation obtained a grant from the U.S. Department of Health, Education, and Welfare to hold a meeting called the "Enigma of Central Nervous Regeneration."[6] In February of 1970, leaders gathered in Palm Beach Florida to review progress and to outline strategies. Given the startling experimental results, the conferees reached conclusions that represented a sea change in attitudes toward regrowth and regeneration: "The problem of regeneration in the central nervous system, previously considered hopeless, is amenable to solution through basic research . . . required to achieve the breakthrough. It is now entirely reasonable to abandon the view that central nervous system regeneration cannot be accomplished in man."[6]

Regrowth in Sweden

With visions of the stricken arising from wheelchairs and of stroke victims taking their first few steps, an entirely new line of experimentation was unfolding: brain transplantation. The scene shifts from Palm Beach to Sweden, ever at the frontier of neuroscience. Not one, but two scientific teams were planning experiments on the near-phantasmagorical project of transplanting neurons to the central nervous system. Even in the new-found optimism of the moon-shot 1970s, the approach seemed a wishful hybrid, one part beneficent Frankenstein, one part Rube Goldberg, and one part lunacy.

Lars Olson at the Karolinska Institute in Stockholm and Anders Bjorklund at the University of Lund worked in different worlds in Sweden. The Karolinska is the academic-scientific jewel of international, urbane Stockholm. Lund is a pastoral, medieval, university city of tree-lined boulevards. Bjorklund of Lund is a compact, intense, focused inquirer who overlooks no detail in his work or personal interactions. He is short, rather formal, crisp in speech, and authoritative. He moves with an unmistakable air of command. Though never overbearing, his opinions are definitive, clearly articulated, and deeply perspicacious.

Lars Olson of Stockholm, by contrast, projects a reserved, nearly casual, image. Tallish, sandy haired, blue-eyed, and informal, he can usually be spotted draped in a backpack. Olson commutes to the laboratory on a decidedly low-tech bicycle that some suspect is of nineteenth-century vintage. He lives with his family on one of the islands in the Stockholm archipelago. He allegedly sleeps in his grandfather's (or is it his great-grandfather's?) bed. His statements frequently sound like questions, and his strongest assertions often have the ring of musings. These externals surround a profound thoughtfulness—a creativity that is constantly searching for the implications and applications of his scientific insights.

Their contrasting personal styles clothe two of the most extravagantly productive and imaginative neuroscientists of the late twentieth century. They have been at the forefront of enough scientific discoveries to maintain colleagues in a state of constant breathlessness. Since they were hardly unaware of each other, Swedish myth has it that there may even have been a competitive element in the mix.

Both Bjorklund and Olson studied brain development. Both used the new (Swedish) methods for investigating neurotransmitter fluorescence in the brain. Both were anatomists and microscopists. Both had discovered in 1973–1974 that neural pathways descending from the brain into the spinal cord could regrow after surgical cutting. Olson's team had even showed that the regrowth was accompanied by recovery of hindlimb motor reflexes, a phenomenon that held out the possibility of true recovery of function. The stage was set for the next leap.

To explore the growth capacity of neurons in the adult brain, to clearly distinguish the growing neurons from their surroundings, to anticipate manipulating the growing neurons in the future without disturbing the surrounding brain, to dream of functional recovery and a brave new world of treatments, a novel experimental approach was needed. Lightning struck in Lund.

Bjorklund's group published their findings in *Nature* in 1976.[7] From opening salvo to last paragraph, it reads like a series of banner headlines. A few words from the title capture the excitement: "Growth of transplanted . . . neurons into the adult hippocampus . . ."[7]

Wasting no space, the team jumps into a provocative summary statement of promise in the first paragraph: They reported that transplants grow into the partly denervated hippocampus, demonstrating remarkable regeneration.[7]

Bjorklund's team had transplanted small pieces of embryonic brain into the adult and detected robust growth of the donor neurons into the host hippocampus, the memory center. The age of brain transplantation had begun. The potential for the future is articulated in the last paragraph. One of the next opportunities, and challenges, was incorporated in the concept that the brain contains intrinsic mechanisms that foster reconnection. The challenge was to harness this capacity. Olson in Stockholm followed an exotic path to brain transplantation. As a graduate student at the Karolinska, his doctoral project involved a massive study of neuronal growth in rats. While the conventional approach usually consisted of observing nerve growth microscopically in developing animals and describing the maturing nervous system, Olson employed his own unique twist. He was driven by a passion to know how individual, defined neurons developed under different conditions in the living animal. How could neurons be exposed to different conditions in tiny rat pups without destroying both neurons and rat? Olson perfected a typically (for Olson) novel system.

As a young graduate student, he transplanted brain neurons from developing rats to the eyes of adult rats.[8] The eye contains a tiny, fluid-filled chamber in front of the iris. The chamber is enclosed by the cornea in front. Using the most delicate of needles, Olson introduced nerve cells into the chamber lying between the clear cornea in front and the iris behind. The chamber acted like a miniature culture, allowing the donor neurons to develop in the live animal. Moreover, since the cornea is transparent, the growing neurons can be observed directly in the living rat. Chemicals or other pieces of tissue were introduced to examine their effects on the growing neurons in the living animal.

With his remarkable experimental paradigm, Olson attacked a welter of critical questions in developmental neuroscience that had been unapproachable. Do different neurons grow differently in the same environment in the living rat? Are growth requirements therefore different for different neurons? Are neurons intrinsically diverse? By implication, would different neurons demand different treatments to induce regrowth? Do neurons have different growth requirements at different stages? The important, now tractable, questions stretched on and on. Olson's system allowed him to ask questions that were formerly unapproachable in the living animal. Driven by his intensity, he formulated and began addressing an astounding number of these problems. In the course of little more than a Stockholm summer of midnight sun, he worked round-the-clock, examining over five thousand transplantation specimens. Discoveries poured forth almost daily.

One example illustrates how Olson reoriented our thinking about neurons, growth, and regrowth. How do neurons know to innervate the correct targets, and how do neurons know to form the correct pattern of innervation for diverse, individual targets, all of which lead to proper cell-to-cell connections? These questions could scarcely be of greater importance in the nervous system, which depends on precise circuitry for functional communication. To approach these questions, Olson cotransplanted small pieces of different targets with the same types of neurons, and different neurons with the same targets in other experiments. In a simple but elegant tour de force of experiments, he found that it is the targets that dictate which neurons innervate them and how the pattern of innervation by the neuron fibers form. The targets contain specific information that directs innervation by neurons. Individual neurons are not burdened with a need to store information about a myriad of individual targets they might encounter. In answer to a perennial question of how neurons choose the correct targets, it is the targets that confer specificity and selectivity on innervating neurons. Thus targets hold at least some keys to the formation of the nervous system's complex circuitry. Though not explicitly appreciated at the time, this would constitute evidence that targets elaborate growth factors.

While Olson's scientific discoveries provided glorious insights into growth processes, the implications for brain disease were at least as profound: nerves could be transplanted to the living. The nerves grew, projected to correct targets, and innervated individual targets normally.

Regrowth, Transplantation, and Brain Disease

Working a few miles apart, employing their own novel approaches, Bjorklund and Olson[9] had moved neuroscience into the new world of transplantation. Though the promise has yet to be fully realized in human disease, we find that now, twenty-five years later, the very concept of brain disorder has been utterly transformed. The unthinkable treatment is now a rational topic of conversation. Although transplantation could affect the fate of people with disorders as diverse as stroke, brain trauma, and birth defects, its promise is most clear-cut in Parkinson's disease.

Parkinson's disease has plagued man through the ages. As we have seen, the disease is caused by the death of nigral neurons deep in the brain, neurons that govern the "motor programs" that coordinate complex movements. The relentlessly progressive illness strikes men and women in middle age and late life. The features of Jeanette McCready's decline are typical: slowed movement, muscle rigidity, impaired gait, tremors, and a masklike face. As the disease progresses over months to years, weakness often leads to near-paralysis that condemns its victim to a life in bed and wheelchair. Ultimately, the most severely afflicted require constant nursing care in which they are fed, bathed, washed, and assisted with their toilet needs.

The experimental destruction of the nigrostriatal system in rats or monkeys leads to Parkinson's disease, and provides an animal model allowing potential treatments to be tested. The Bjorklund and Olson teams used this model to pioneer the use of neural transplantation. While it is customary to describe scientific results in a cold, dispassionate, objective manner, their breakthroughs are miraculous.

Though employing different tactics, their basic strategies were similar. Each team destroyed the Parkinson's neurons in rats. When the rats developed the signs and symptoms characteristic of Parkinson's, the teams then transplanted the appropriate neurons into their brains to replace the ones that had been destroyed.

A new era in the treatment of hopeless neurologic disease began. Transplantation markedly improved the disease in the experimental animals. Their weakness and paralysis were transformed into virtually normal movement. Immobile animals began walking. Rigid rats became flexible and dexterous again. The experiments were repeated in our closer relatives, monkeys, with the same results. If anything, the recovery of the monkeys was even more dramatic because their disease so closely mimicked human Parkinson's. Experimental transplantation has already begun in humans. Will it work for Alzheimer's?

On its surface Alzheimer's, with its manifold mental deficits, would seem to be a poor candidate for transplantation therapy. But the discovery that the degeneration of neurons at the base of the brain ("basal forebrain neurons") is associated with memory loss has raised intriguing possibilities. The dying neurons serve certain types of memories, especially spatial memory. Without them, we cannot make our way from work to home or bedroom to bathroom. We get lost. In a similar fashion the destruction of these neurons in rats is devastating. The rats no longer can remember routes through mazes and become hopelessly lost. Again, the transplant approach could be tested in an animal.

Bjorklund embarked on groundbreaking experiments with Rusty Gage, an American working in his laboratory. Gage, a U.S. expatriate raised in Rome, has a family tree with a deeply ironic relation to the mind sciences. His great grandfather, Phineas Gage, had unwittingly contributed to neuroscience while building the transcontinental U.S. railway. A railway spike had been accidentally driven through his brain, changing his personality from upstanding, salt-of-the-earth worker to ne'er-do-well reprobate, thereby contributing to our understanding of brain-mind relations (see Chapter 6 for details).

Gage and Bjorklund approached Alzheimer's and transplantation with several ingenious paradigms. In one approach, they used no tricky assumptions whatsoever.[10] They simply tested a large group of aged rats (2 years and older). They picked out the population of dunces by carrying out a maze-learning test and worked with those animals. The question was simple. Could maze learning be improved by transplanting the basal forebrain memory neurons? The answer was startling. Transplantation dramatically enhanced maze learning in the aged, demented rats. In parallel experiments they purposely damaged the rats' basal forebrain neurons. Again, transplantation replacement restored spatial memory to almost normal.

Bjorklund and Gage had demonstrated that, in principle, transplantation was a viable approach to the recovery of function in at least one type of dementia. With a little help, the brain might, after all, be a renewable resource.

In their experiments Olson and Bjorklund/Gage had transplanted neurons to damaged areas of the brain. In the normal brain, though, neurons send long fibers that act over great distances. Could action-at-a-distance recover after illness and injury? This would require the regrowth of long fibers to target neurons—yet another piece of the puzzle of regeneration and the recovery of function after brain disease.

This was the very problem that Albert Aguayo[11] was attacking at the Montreal General Hospital. Aguayo, a clinical neurologist, was one of the generation of intellectuals who fled Argentina during the dark period of military dictatorship. Members of this group emigrated to North America and Europe, and emerged as scientific leaders in academia and industry. Aguayo had been experimenting with peripheral nerves in his studies of regrowth after injury, and discovered that the highways formed by non-neural support cells were critical in directing the orderly regrowth of peripheral nerves. These support cells formed tubes or tunnels through which a spectrum of peripheral neurons regrew. Aguayo's experiments indicated that the support cell highways were more important than the neurons themselves in fostering regrowth of the neuronal fibers. He reasoned that peripheral support cells might encourage reluctant brain neurons to grow as well. Perhaps the inability of central neurons to regrow long processes was due to problems in the cellular environment and not to intrinsic deficits in the neurons. Cajal had suggested as much nearly a century before.

Armed with this hypothesis, Aguayo's team used peripheral nerve bridges to link distant areas of the central nervous system. They used the normal peripheral nerve tubes of the sciatic nerve, which consists solely of support cells. One end of the sciatic nerve bridge was inserted into the brain just above a rat's spinal cord. The other end was run outside of the spinal cord under the skin and then inserted into the spinal cord in the chest. The entire length of the bridges was about 6 inches, roughly as long as the rat's body. The results were crystal clear. Nerve fibers from the brain grew through the bridge into the spinal cord. In turn, nerve fibers grew in the reverse direction from the spinal cord through the bridge into the brain. To determine whether other neurons in the central nervous system also possessed this growth potential, Aguayo used nerve bridges in the cerebral hemispheres. Nerve fibers grew through the bridges in these higher brain centers, thus confirming the spinal results. In one particularly provocative experiment, the neurons known to die in Parkinson's disease grew vigorously through the bridges to their targets.

The conclusion of his resulting scientific article is a ringing affirmation of the emerging insights: the growth of central fibers is governed by interactions with surrounding cells; failed regeneration is not due to an intrinsic lack of regenerative capacity.[11]

It is time to reconnoiter. From a deep well of hopelessness a few short decades ago, we have progressed to guarded optimism. Why? Because we now know that nerves adjacent to an area of damage can sprout new fibers in the brain and the spinal cord. The nerve fibers make synaptic connections that could lead to communication and even recovery. Neurons transplanted to the brain of old or damaged animals do regrow and lead to recovery of function. What is more, rat Parkinson's disease exhibits striking improvement after transplantation. Rats with brain damage and deranged spatial memory are dramatically better after brain neuron transplantation. Peripheral nerve bridges encourage brain and spinal neurons to grow impressive distances to targets, potentially allowing access to even the most delicate brain areas.

Do these striking advances point the way to new therapies? And how do these insights change our concepts of the brain and its diseases? One type of treatment already under exploration is that of cell and gene therapy. In principle, any cells, not just neurons, can be transplanted to the ailing brain. Some cells, which elaborate growth and survival factors better than do neurons, may be more effective in treating Alzheimer's or Parkinson's diseases. We are certainly not restricted to the use of brain cells in transplantation. With other cellular populations, we may be able to treat stroke, head trauma, and degenerative disease far more effectively. By judiciously choosing specialized cells within the body, we could devise optimal transplants customized for particular diseases in individual patients. We can even dream of rebuilding a diseased brain with cellular parts that are more effective than the original.

Our dreams may take us even farther. In this age of molecular biology, we are not restricted to cells naturally occurring in the body. We can now design new cells for new functions by relying on genetic engineering. For example, in Parkinson's disease we can design cells by inserting new genes that make the transmitter signal which controls motor function; also, we can potentially insert genes that make survival signals that prevent the degeneration of the dying neurons. Similarly, for Alzheimer's disease we could insert genes that replace defective ones in hereditary forms of the disorder. Though all this progress still dwells in the realm of speculation, the potential is stunning.

Do these discoveries alter fundamental concepts of mind and brain? While we might ultimately leave this question to the philosophers, some modest observations may be in order. The brain is no longer a locked, inaccessible, inviolable box. Illness and injury need no longer be regarded as irreparable tragedies that forever doom its protagonists to dehumanization. If not truly renewable resources, brain and mind are therapeutically approachable. Brain and mind are not immutable switchboards, but rather, flexible miracles of nature amenable to fruitful interactions with the environment.

EPILOGUE:
ENOCH'S ENDGAME

Enoch entered Broome Senior Care Home on a cloudy, humid July third, an ironic preamble to Independence Day. Sally couldn't decide whether the transition signaled a new phase of independence for her, for Enoch or for both. The last six months had been a disorienting, downward nightmare.

After extreme effort to hold the pieces of Enoch's behavior together, everything seemed to fall apart at once, and Sally could no longer cope. Miss Parties, the visiting nurse, was wonderfully efficient and gentle, but Enoch's confusion was overwhelming.

Imperceptibly but rapidly, his confusion spread until it invaded his every act, every experience. Though Sally could not fix a specific denouement date, Enoch could no longer reliably find his way around the apartment, day or night. He was an alien wanderer in his own house. The momentary lapses were replaced by a constant need for guidance.

Miss Parties had to escort Enoch to the bathroom, and then back to the living room. Once guided to the kitchen, he could not be left alone. At one point, he lit the gas oven top with a dishtowel over the burner. On another occasion, he left the tap running and the kitchen flooded. Knives were scattered, often hidden dangerously, the microwave was left on, sponges appeared in the dishwasher, the freezer remained open. With each misadventure, he was able to understand the error, but the information had no effect whatsoever on the next miscue. Sally was living on the edge of disaster. The sense of imminent catastrophe was constant.

Personal hygiene and toiletry disintegrated. Enoch was routinely incontinent of urine in bed at night, and wore a modified diaper. He was even unaware of his incontinence. His indifferent behavior was a source of shame and grief for Sally. Enoch often urinated in a corner of the bed-room, seemingly totally indifferent to the social impropriety of the act. Even this behavioral symptom was well recognized by Dr. Rudick, the neurologist, as a strange, but common sign of frontal lobe dysfunction. It was graced by the name, "frontal lobe bladder syndrome," explained the neurologist in his antiseptic tone. Names and explanations did little to re-lieve the now-constant stench that permeated their home. "Thank God," thought Sally, there was no bowel problem; but this comforting thought was a source of deep foreboding about difficulties to come. She had lost a life-long ally: her faith in the future. The optimism that had sus-tained her through vicissitudes and crises since childhood evaporated un-der the onslaught of Enoch's deterioration.

Enoch's vegetative disintegration, his incontinence, his spatial and temporal disorientation, and his carelessness paled by comparison to the fragmentation of his mental life. He misplaced the newspaper, that he no longer comprehended, lost glasses, pen, pencils, coffee cups, towels, arti-facts of a life no longer lived. Despite frequent visits, he remembered the faces of his business partners but was at a loss to recall their names. While the visitors were embarrassed, Enoch acted only mildly distracted. Sally was devastated when she finally accepted the fact that he could not reli-ably recall the names of their children or grandchildren. Of this, he was acutely aware, but was accepting in an almost alien, objective manner.

At Broome, nursing supervision eliminated the dangers of the home, but Enoch deteriorated even more rapidly. He now urinated in the hall-way, in the small cafeteria, and in a corner of his room. He needed help eating his meals, and began gagging on his food. Sometimes he held the food in his mouth, while at others, he attempted to swallow an unchewed mouthful. On one occasion, he nearly choked, became short of breath, fi-nally coughed it up, but was left with a mild pneumonia, which resolved after two weeks of antibiotics and pulmonary therapy. The bout left him 11 pounds lighter, with a wasted, debilitated physical carriage.

Whether due to his weakened state, his underlying disease, or coinci-dence, a series of physical problems followed. In spite of his incontinence, he experienced several episodes of inability to urinate, though his bladder was full. He required bladder catheterization, passage of a tube, but this left a urinary tract infection in its wake. With chills and fever, he lost his appe-tite completely, became indifferent and nearly mute. The infection was eliminated by sulfa drugs in three weeks, but he never returned even to his former compromised state. Enoch was alert, but distracted most of the time, and efforts had to be made by the staff to attract and hold his attention. He remained pleasant, but exhibited a docility and passivity that prevented him from initiating any activity on his own.

On clinical rounds one morning the staff physician noted that Enoch was covered with small black-and-blue bruises. Blood tests revealed no clotting deficiency, but did detect moderately severe iron deficiency anemia. No underlying cause could be diagnosed, and the condition was attributed to his general debilitated state and a now chronic, low-grade urinary tract infection.

Enoch grew increasingly listless and unresponsive, and tended to stay in bed most of the day, unless actively mobilized by the staff. In spite of his inactivity, he experienced bouts of mild shortness of breath, required two pillows to sleep in comfort, and developed ankle swelling that varied from day to day. The doctor prescribed a diuretic and a digitalis preparation, and although the shortness of breath improved, he experienced skipped heartbeats that he found frightening. Through it all, confusion, disorientation, and distraction continued to increase the distance between Enoch and the world around him. At times he was barely in touch, though he seemed to be alert.

The contrast of five years ago between his mental debility and his apparently normal, vigorous, physical well-being was all but gone. Then he was a forgetful, distractible, intermittently confused, but normal appearing lawyer. His physical state now reflected profound dementia. Enoch's distant, unfocused gaze, his disheveled hospital gown, the brown stains on his pajama collar, his hair projecting at odd angles, even his chapped, cracked lips and cold sores, were signposts of a world his friends and family could not enter. That world was increasingly the universe of his childhood. Enoch spent a great deal of time talking to the attendants about his beloved uncle Al. Enoch had loved animals, and Al took the eight-year-old Enoch to the Bronx Zoo, his second home, almost every weekend. Enoch ducked under the wooden turnstile, mirror-polished by thousands of hands, filled with explosive excitement that propelled him up the stone stairway to the llama enclosure, on the way to the African Great Plains. Though he and Al always stopped to feed the llamas ("keep your hand flat so they can't bite"), it was Africa and the plains that transported the boy to the other world. The Plains stretched on for acres, with scattered trees, just like the photos of the veldt that Enoch devoured at home. The great expanse was all visible, since it was at a level 20 feet below the surrounding walkway. Enoch marveled at the distances, but it was the animals themselves that carried him to Africa. The din of the calling, displaying peacocks, the elegant, fastidious gait of the crowned cranes, the weightless flights of the gazelles, and even the omnipresent flocks of New York pigeons announced arrival in the enchanted land.

Enoch's attempts to describe the peacocks and cranes and gazelles to the attendants at Broome were met with blank confusion. Even Molly, who always made the bright morning bed with a friendly, singing West Indian air, shook her head with a tolerant, disbelieving smile. Thursday evening, Enoch was mildly surprised to see uncle Al come into the room. They began talking about the zoo and the African Plains. Al seemed to get bored, and in a state of some confusion, Enoch ended up talking to his new friend Gus, one of the Broome night attendants. Gus smiled his warm, wonderful smile, "You was a kid then, Mr. Wallace, a long time ago. Those 'a good times to hold onto, Mr. Wallace, good times." Gus didn't know about any Plains, and couldn't rightly recall whether "'zels and cranes" were real or make believe, but he knew good times for certain, and was happy with his friend.

Enoch's associates and family were another matter. At first, his partners visited as a group once a week. But as Enoch talked more and more about the zoo and uncle Al and the plains, the visits became less frequent. Enoch and Sally's friends, though a diverse lot, visited with a shared funereal sense. They talked to Enoch as if he was no longer there. Though he didn't seem to notice, Sally was always shaken as Enoch was treated like a ghost. Even the irrepressible Mina was subdued in his presence, her wit and fire extinguished by Enoch's misty ramblings. The kids visited dutifully, listening tolerantly as Enoch drifted from the present to the ever more remote past, but they gradually focused on details of home, their children's school, piano and soccer practices, and work schedules that failed to attract Enoch's attention. Of course, the grandchildren were not allowed to visit Broome for fear of importing an exotic, deadly contagion from the dark suburbs that would lay waste to patients and staff.

Sally was numb as Enoch drifted in and out of childhood. His attention to the present now focused on the immediate details of life: the soup on the tray, the gelatin dessert, the cut of his slippers and the soap in the bathroom formed the borders of his world. Sally was unable to penetrate to anything much more substantive, but, in truth, had really given up trying.

Enoch's Alzheimer's disease was pursuing a gradual course.

AFTERWORD:
LEST YE SEE MIRACLES

Who are we? Where did we come from? Instead of answers, we have tentative impressions. Though we do not yet know who we are, we know who we are not. We are not static, inflexible switchboards; we are not mute, inanimate computers; we are not separate from the world around us. We are dynamic, ever-changing individuals, integral parts of the nature in and around us.

The central paradoxes that characterize our nature contribute to the difficulties of self-definition. Mind and brain exhibit astounding flexibility, mutability, and plasticity in the face of the ongoing stability that marks our ontological continuity. Change and constancy interact, confounding attempts to explain cognition, consciousness, awareness, and emotion. However, plasticity allows approaches to madness, epileptic possession, and idiocy, that were inconceivable a few short years ago.

Plasticity expresses the unity of our internal and external worlds, the unity of organism and environment, the unity of nature and nurture. Paradoxically, biology maintains an individual's separate identity simultaneously. The apparent contradiction of unity and identity further complicates efforts to grasp our essence. Yet the Promethean insight that environment regulates brain which regulates environment is emancipating. It provides the therapeutic keys to free us from the most devastating disorders of mind.

The multiplicity of structural levels in brain and mind further complicates comprehension. Genes, molecules, cells, systems, multisystems, and brain regions interact to generate cognition, consciousness, and emotion. The explanatory language for one functional level is inappropriate for another. While one goal of neuroscience is to integrate levels, causally linking brain and mind, we have yet to discover unifying descriptors. Yet the realization that multiple levels drive function now offers diverse, novel loci for treatment of mind disease. Gene therapy, brain cell transplants, conventional drug treatments, and psychotherapy are being successfully applied at different organizational levels in the very same mind disorders.

Though we have not solved the mind-brain problem, our limited triumphs along the way have revolutionized approaches to neuropsychiatric disease. The revolution has depended on the discoveries that the brain is changeable, or plastic, that unity with the environment allows change from the outside, that the rules by which structure governs function can be decoded.

What is life like in the new world of neurology, psychiatry, and psychosomatics? How has neuroscience revolutionized our approaches to Enoch Wallace, Charlene Washington, Jeanette McCready, Andre Mallard, Jeremy Soreno, and Elias Meharry? What is actually different?

To gain perspective, we list a number of seemingly unrelated discoveries that form the basis of the revolution:

Specific populations of neurons in the brain control specific behaviors and discrete mental functions.

The brain is home to myriad growth and survival factors that govern the life and function of specific cell populations.

Identified, cloned genes control selected mental abilities and behaviors.

Specific environmental stimuli, whether physical, verbal, or situational, activate specific brain genes, expressing plasticity.

Contrary to received wisdom, neurons are actually produced in the adult brain.

Specific experiences increase the proliferation of reproducing brain neurons.

These discoveries comprise a blueprint for developing new treatments for formerly hopeless neurologic diseases. We can illustrate the new neurology by applying these insights to a model degenerative disorder, Parkinson's disease. The population of neurons that progressively dies in Parkinson's, deranging motor programs, has been identified.[1] This knowledge has opened the way for cell replacement therapy. Transplantation of replacement cells into the brains of mice, rats, and primates with experimental Parkinson's dramatically improves the signs and symptoms.[2] The era of experimental cell therapy has begun.

Intensive worldwide searches are now underway to identify the best cells for replacement. Brain cell populations, cells from the closely related adrenal gland, stem cells, and genetically engineered cells have already been used. Painstaking, step-by-step modification of the cells is now resulting in ever-improving therapeutic responses. Experimental cell therapy has been used in a few end-stage patients; responses are not clear cut. Nevertheless, we are systematically closing in on a cell replacement bedside approach to this devastating disease.

The discovery of the neuronal population that dies in Parkinson's led to other critical questions and advances. Are the "Parkinson's neurons" responsive to growth and trophic factors, like other brain populations? Years of exhaustive experimentation ultimately led to the discovery of a number of such factors. The most potent to date is termed "glial cell line derived neurotrophic factor" (GDNF).[3] GDNF markedly enhances survival and growth of the Parkinson's neurons in cell culture. Injection of GDNF into the brains of animals with experimental Parkinson's results in dramatic improvement. GDNF has already been administered to Parkinson's patients on an experimental basis. While results have not yet been compelling, a multitude of conditions may have to be optimized before satisfactory responses are possible. The age of factor therapy for Parkinson's has begun.

If we can replace cells and help Parkinson's, can we replace genes? Which genes? How would we recognize a Parkinson's gene, if we found one? What would a Parkinson's gene even look like? Where would a search begin? Let's return to first principles. We have identified the cells that degenerate in the disorder. We know one critical signal that they send is dopamine. We have been able to treat the disease by replenishing the depleted signal with L-DOPA. Can we use gene therapy instead of L-DOPA therapy to replace dopamine? Does this question make any sense? In theory, we could use genes that produce the enzyme machinery that synthesizes dopamine. This strategy has been tried. Cells were transfected with the gene (the gene was inserted) that produces the key dopamine synthetic enzyme.[4] The cells were then transplanted into the brains of animals suffering from experimental Parkinson's. Only a fraction of the cells survived in the brain after transplantation. The brain was examined for dopamine production. The cells that survived did synthesize the signal. The animals exhibited dramatic improvement. Gene therapy can work in Parkinson's disease, at least in experimental Parkinson's in rats.

While this work is a first step on the gene therapy path, dopamine replacement does not prevent death of the "Parkinson's neurons." Can we identify any defective genes that underlie Parkinson's cell death? Recently, a gene defect has been discovered in a family with a rare form of the disease.[5] Although this mutation can account for only a tiny fraction of patients with the disease worldwide, it may provide hints about underlying causes. Replacement of the defective gene may comprise a new gene therapy approach, possibly preventing the disease symptoms from ever appearing in these patients. But replacement gene therapy is in the dreaming-thinking stage, and has not yet begun.

Transplantation of genes, transplantation of cells. Recent shocking discoveries may render transplantation old-fashioned. Contrary to hallowed dogma, cells can reproduce in the adult brain and give rise to new neurons. Neural "stem cells" have been discovered in the brain (Chapters 6 and 10),[6] which means that the brain may be able to replace its own degenerating neurons. If proliferation could be triggered in the brain, in a population-specific fashion, perhaps the dying Parkinson's neurons could be replaced from within. How could we encourage the birth of new neurons in the brain?

What new treatment, what exotic medication, what electromagnetic frequency could unleash proliferation in the living brain? Available answers are more bizarre, and more mundane, than could have been imagined. One answer seems to be "experience."[7] An enriched environment and physical activity increase proliferation of brain cells in living rats. If the rats are housed with toys, games, and diversions, more cells are produced and survive in their brains. If the rats are allowed to run on a treadmill, more cells are produced. Nightly exercise (for the nocturnal rat) in a fun-filled home increases the number of new neurons. Daily exercise in the amusement park may banish our quaint twentieth-century high-tech approaches to the junk heap of peculiar artifacts. Neuroscientists await the judgment of history.

The neuroscientific revolutions have delivered L-DOPA neurotransmitter drug therapy, cell population therapy, growth factor therapy, trophic factor therapy, cell transplantation, gene therapy, neural stem cell proliferation therapy, and environmental-experiential-situational therapy for use in Parkinson's. Neuroscience is creating potentially new treatments faster than they can conceivably be applied. The number of new therapies may outstrip the number of patients available for testing.

To appreciate the scope of the neuroscientific revolution we turn from our focus on the Parkinson's example to an overview of brain and mind disease as we enter the new millennium. Let's undertake the impossible task of summarizing advances of the past ten years. Though our list will be hopelessly inadequate, we may capture the flavor of health and disease in the next century.

We begin with the Holy Grail and Godhead of the late twentieth century, the gene. In a frenzy of discovery, genes have been cloned for an astounding array of disorders, including the present plagues of mankind. Genes for Alzheimer's, Huntington's, Parkinson's, and Lou Gehrig's diseases, for Fragile X mental retardation syndrome and forms of epilepsy have been isolated. Genes for receptors, transmitters, and electrically conducting ion channels have been cloned. These discoveries are revealing the innermost workings of the brain, and are indicating new therapeutic approaches to seizures, nerve-muscle disease, and madness.

Miraculously, however, these insights are nearly empty compared to a new revelation. Neuroscience is now defining the physical basis of pleasure and pain. Genes for pain signals and receptors have been cloned.[8] Pain itself may be an experience of the past. With an emerging understanding of its physical basis, we are now able to eliminate major forms of pain. We are rapidly approaching a time when we can banish pain itself. What will life be like without pain? The very nature of human experience is subject to reexamination. What does this mean for our conception of the joys of Heaven and the pain of Hell? Can we survive in a world without pain?

Neuroscience is also storming the painful medieval dungeon of madness. Neuropharmacology has revealed transmitter mechanisms associated with the scourge of depression. The product, Prozac, has delivered a new life of freedom to millions.[9] We contemplate a life free of suicide, despair, and self-hate.

Prozac has been emancipating in another, wholly unanticipated fashion. A tragic group of individuals has been enslaved by their own uncontrollable behaviors, repeating acts against their will. The victims have been held guilty for their behavior, and have been subject to mockery and abuse. Prozac magically eliminated the imprisoning behaviors, revealing disease instead of depraved deviance. A medicine cured a disease and lifted the curse of superstition. The possessed patients were suffering from obsessive compulsive disorder; they were not bewitched. Prozac ended their dark night of Inquisition.[9]

"Patients never recover from spinal cord injury." Though tragic, this neurological North Star provided grim orientation for neuroscientists from time immemorial. That certainty has changed utterly. Growth factors and inhibitory factors have been discovered in the spinal cord. Paralyzed rats take a few tentative steps after treatment with antagonists of inhibitors and with growth factors. Spinal nerves actually regrow.[10] Transplantation of stem cells to the injured cord also fosters recovery. Although the Rubicon has been crossed, these breakthroughs are pedestrian next to a revolutionary treatment now employed in humans.

After severe injury, the spinal cord is separated from the brain, causing complete paralysis. Impulses cannot travel from the brain's voluntary movement centers to the cord. Isolated from the brain, the spine is worthless, drifting flotsam, a mockery beneath the quadriplegic's open bedsores. Movement is a fading memory occurring only in desperate, suicidal nightmares. A neuroscientific discovery has now fulfilled the ancient wish and prayer of Lourdes. The devastated, mortally wounded cord, the ragged fragment of flotsam, can be *taught* to move again. After years of painful, meticulous physical training, patients detect a flicker of muscle movement. The flicker imperceptibly becomes a real movement. Paralytics are standing at the dawn of an unimaginable miracle.[11] This is the triumph of the will, the triumph of neural plasticity.

"Lest ye *see* miracles, ye shall not believe." All of these discoveries, though exciting and revolutionary, involve abstractions that lie beyond visualization. What does a neurotransmitter look like? How do we picture a receptor? For that matter, how do we image a thought, an emotion, attention? Absurd religious question. But neuroscience is providing the answer. MRI (magnetic resonance imaging) and PET (positron emission tomography) have pictured the brain systems underlying specific thoughts, emotions, memory, and attention.[12] We can actually *see* brain dysfunction in schizophrenia. Lest ye see miracles.

NOTES

Chapter 1

1. Bliss, T.V.P. and Lomo, T.J. *Physiol. Lond.*, 232, 331–356, 1973. Morris, R.G.M., Anderson, E., Lynch, G.S., and Baudry, M., *Nature* 319, 774–776, 1986.

2. For background and overview see Levi-Montalcini, R. "NGF: An Uncharted Route," in *The Neurosciences: Paths of Discovery*, pp. 245–265, MIT Press, 1975.

3. Levi-Montalcini, R., *In Praise of Imperfection: My Life and Work,* Basic Books, 1988; Levi-Montalcini, *Sci. Am.* 68–77, 1979. For the definitive collection of research articles see Levi-Montalcini, R., *The Saga of the Nerve Growth Factor,* World Scientific, Singapore, 1997.

4. Honneger, P. and Lenoir, D., *Dev. Br. Res.* 3, 229–238, 1982; Gnahn, H., Hefti, F., Heumann, R., Schwab, M.E., and Thoenen, H., *Br. Res.* 285, 45–52, 1983; Hefti, F., Dravid, A., and Hartikka, J., *Br. Res.* 293, 305–311, 1984; Hefti, F., Hartikka, J., Eckenstein, F., Gnahn, H., Heumann, R., and Schwab, M. *Neuroscience* 14, 55–68, 1985; Martinez, H.J., Dreyfus, C.F., Jonakait, G.M., and Black, I.B., *Br. Res.* 412, 295–301, 1987; Mobley, W.C., Rutkowski, J.L., Tenekkoon, G.I., Buchanan, K., and Johnston, M.V., *Science* 229, 284–287, 1985; Seiler, M. and Schwab, M., *Br. Res.* 300, 33–39, 1984; Shelton, D.L. and Reichardt, L.F., *PNAS USA* 83, 2714–2718, 1986; Whittemore, S.R., Ebendal, T., Larkfors, L., Olson, L., Seiger, A., Stromberg, I., And Persson, H., *PNAS USA* 83, 817–821, 1986.

5. For reviews see Barde, Y-A., *Neuron* 2, 1525–1534, 1989; Chao, M.V., *Neuron* 9, 583–593, 1992.

6. For reviews see Lo, D.C., *Neuron* 15, 971–981, 1995; Black, I.B., *J. Neurobiol.* 41, 108–118, 1999.

7. Skaggs, W.E. and McNaughton, B.L., *Science* 271, 1870–1873, 1996.

8. For positive effects of stress on neuronal communication and learning, see Shors, T.J. and Dryver, E., *Br. Res.* 666, 232–238, 1994; Shors, T.J. and Servatius, R.J., *NeuroReport* 6, 677–680, 1995. However, some forms of stress may be deleterious to certain forms of learning, as reported in Shors, T.J. and Dryver, E., *Psychobiology* 4, 247–253, 1992.

9. For a classic book derived from a doctoral thesis, see Iversen, L.L., *The Uptake and Storage of Noradrenaline in Sympathetic Nerves,* Cambridge University Press, Cambridge, U.K., 1967.

10. For review see Hendry, I.A., "Control in the Development of the Vertebrate Sympathetic Nervous System," in *Reviews of Neuroscience,* S. Ehrenpreis and I.J. Kopin, eds., pp. 149–194, Raven Press, New York, 1976. For original reports see Hendry, I.A., *Br. Res.* 134, 213–223, 1977; Hendry, I.A., Stach, R. and Herrup, K., *Br. Res.* 82, 117–128, 1974; Hendry, I.A., Stockel, K., Thoenen, H., and Iversen, L.L. *Br. Res.* 68, 103–121, 1974.

Chapter 2

1. For review see Greene, L.A. and Shooter, E.M., *Ann. Rev. Neurosci.* 3, 353–402, 1980.

2. Chao, M.V., *Neuron* 9, 583–593, 1992.

3. Greene, L.A. and Tischler, L.A., *PNAS USA* 73, 2424–2428, 1976; italicized term added.

4. For review see Kaplan, D.R. and Stephens, R.M. *J. Neurobiol.* 25, 1404–1417, 1994.

Chapter 3

1. The classical studies of Axelrod, Carlsson, and von Euler are cited, summarized, and discussed in Iversen, L.L., *The Uptake and Storage of Noradrenaline in Sympathetic Nerves*, Cambridge University Press, Cambridge, U.K., 1967.

2. Falck, B., Hillarp, N.-A., Thieme, G., and Thorpe, A., *J. Histochem. Cytochem.* 10, 348–354, 1962.

3. Schwab, M.E., Otten, U., Agid, Y., and Thoenen, H., *Br. Res.* 168, 473–483, 1979.

4. For overviews of the societal problem of Alzheimer's disease, see McKhann, G. et al., *Neurology* 34, 939, 1984; Evans, D.A. et al., *J. Am. Med. Assoc.* 262, 2551, 1989; Katzman, R., *Arch. Neurol.* 54, 1201, 1997.

5. Whitehouse, P.J., Price, D.L., Struble, R.G., Clark, A.W., and Coyle, J.T., *Science* 215, 1237–1239, 1982; Coyle, J.T., Price, D.L., and DeLong, M.R., *Science* 219, 1184–1190, 1983.

6. Drachman, D. and Leavitt, J., *Arch. Neurol.* (Chicago) 30, 113, 1974; Drachman, D.A., Noffsinger, B.J., Sahakian, B.J., Kurdziel, S., and Fleming, P., *Neurobiol. Aging* 1, 39, 1980.

7. For review and historical perspective see Bartus, R.T., Dean, R.T. III, Beer, B., and Lippa, A.S., *Science* 217, 408–417, 1982.

8. Martinez, H.J., Dreyfus, C.F., Jonakait, G.M., and Black, I.B. *PNAS USA* 82, 7777–7781, 1985; Martinez, H.J., Dreyfus, C.F., Jonakait, G.M., and Black, I.B., *Br. Res.* 412, 295–301, 1987.

9. Gnahn, H., Hefti, F., Heumann, M.E., Schwab, M.E., and Thoenen, H., *Dev. Br. Res.* 9, 45–52, 1983.

10. Seiler, M. and Schwab, M.E., *Br. Res.* 300, 33–39, 1984.

11. Mobley, W.C., Rutkowski, J.L., Tennekon, G.I., Gemski, J., Buchanan, K., and Johnston, M.W., *Mol. Br. Res.* 1, 53–62, 1986.

12. Shelton, D.L. and Reichardt, L.F., *PNAS USA* 83, 2714–2718, 1986; Whittemore, S.R., Ebendal, T., Larkfors, L., Olson, L., Seiger, A., Stromberg, I., and Persson, H., *PNAS USA* 83, 817–821, 1986.

Chapter 4

1. Morrison, R.S., Sharma, A., De Vellis, J., and Bradshaw, R.A., *PNAS USA* 83, 7.

2. Monard, D., Solomon, F., Rentsch, M., and Gysin, R., *PNAS USA* 70, 1894–1897, 1973.

3. Barde, Y.A., Lindsay, R.M., Monard, D., and Thoenen, H., *Nature* 274, 818, 1978.

4. Barde, Y.-A., Edgar, D., and Thoenen, H., *The EMBO Journal* 1, 549–553, 1982.

5. Leibrock, J., Lottspeich, F., Hohn, A., Hofer, M., Hengerer, B., Masaiakowski, P., Thoenen, H., and Barde, Y.-A., *Nature* 341, 149- 152, 1989.

6. Ernfors, P., Ibanez, C.F., Ebendal, T., Olson, L., and Persson, H., *PNAS USA* 87, 5454–5458, 1990; Maisonpierre, P.C., Belluscio, L.S., Ip, N.Y., Furth, M.E., Lindsay, R.F., and Yancopoulos, G.D., *Science* 247, 1446–1451, 1990; Hollbook, F., Ibanez, C. F., and Persson, H., *Neuron* 6, 845–858, 1991; Berkmeier, L.R., Winslow, J.W., Kaplan, D.R., Nikolics, K., Goeddel, D.V., and Rosenthal, A. *Neuron* 7, 857–866, 1991

7. Jones, K. R. and Reichardt, L.F., *PNAS USA* 87, 8060–8064, 1990.

Chapter 5

1. Glucksmann, A., *Biol. Rev.* 59, 59–86, 1951.

2. Levi-Montalcini, R. "NGF: An Uncharted Route," in *The Neurosciences: Paths of Discovery*, pp. 245–265, MIT Press, 1975; Levi-Montalcini, R., *In Praise of Imperfection: My Life and Work,* Basic Books, 1988.

3. Hamburger, V., *J. Exp. Zool*, 68, 449–494, 1934.

4. Levi-Montalcini, R., *Arch. Biol.* (Liege) 53, 537–545, 1942.

5. Levi-Montalcini, R. and Hamburger, V., *J. Exp. Zool.* 116, 321–361, 1951; *J. Exp. Zool.* 123, 233–287, 1953; *Cancer. Res.* 14, 49–57, 1954.

6. Cowan, W.M., in *Development and Aging in the Nervous System*, pp. 19–41, Academic Press, New York, 1973.

7. Mozer, M.C. and Smolensky, P., Skeletonization: A Technique For Trimming The Fat From A Network Via Relevance Assessment, in *Advances in Neural Information Processing Systems* 1, D.S. Touretsky, ed., Morgan Kaufman Publishers, San Mateo, California, 1989.

8. Kerr, J.F.R., Wyllie, A.H., and Currie, A.R., *J. Cancer* 26, 239, 1972.

9. Johnson, E.M., Jr., *Br. Res.* 141, 105–118, 1978; Johnson, E.M., Jr., Gorin, P.M., Brandeis, L.D., and Pearson, J., *Science* 210, 916–918, 1980; Johnson, E.M., Jr., Rich, K.M., and Yip, H.K., *Trends Neurosci.* 9, 33–37, 1986.

10. For a recent review of this classic work see Martin, D.P., Ito, A., Horigome, K., Lampe, P.A., and Johnson, E.M., Jr., *J. Neurobiol.* 23, 1205–1220, 1992.
11. For an example and a technical introduction to this literature, see Shanahan, S. and Horvitz, H.R., *Genes Dev.* 10, 578, 1996.
12. For a recent review of the Bcl-2 family, see Adams, J.M. and Cory, S., *Science*, 281, 1322–1326, 1998.
13. For overviews of excitotoxicity and apoptosis in disease, with particular reference to stroke, see Choi, D.W., *Trends Neurosci.* 11, 465–469, 1988; Choi, D.W., *Neuron* 1, 623–634, 1988; Choi, D.W., *Cerebrovasc. Brain Metab. Rev.* 2, 105–147, 1990; Choi, D.W., *J. Neurosci.* 10, 2493–2501, 1990.

Chapter 6

1. Rakic, P., *Science* 227, 1054–1056, 1985.
2. Nottebohm, F., *Science* 214, 1368–1370, 1981; Goldman, S.A., and Nottebohm, F., *PNAS USA* 80, 2390–2394, 1983; Paton, J.A., O'Loughlin, B. E., and Nottebohm, F., *J. Neurosci.* 5, 3088–3093, 1985.
3. Altman, J. and Das, G.D., *Nature* 214, 1098–1101, 1967.
4. DiCicco-Bloom, E. and Black, I.B., *PNAS USA* 85, 4066–4070, 1988; DiCicco-Bloom, E. and Black, I.B., *Br. Res.* 491, 403–406, 1989; DiCicco-Bloom, E., Townes-Anderson, E., and Black, I.B., *J. Cell Biol.* 110 (6), 2073–2086, 1990.
5. Wagner, J.P., Black, I.B., and DiCicco-Bloom, E., *J. Neurosci.* 19, 6006–6016, 1999.
6. Richards, L.J., Kilpatrick, T.J., and Bartlett, P.F., *PNAS USA* 89, 8591–8595, 1992; Reynolds, B.A. and Weiss, S., *Science* 255, 1707–1710, 1992.
7. Reynolds, B. and Weiss, S., *Science* 255, 1707–1710, 1992.
8. For reviews see Gage, F.H., Ray, J., and Fisher, L.J., *Annu. Rev. Neurosci.* 18, 159–192, 1995; McKay, R., *Science* 276, 66–71, 1997; Alvarez-Buylla, A. and Temple, S., *J. Neurobiol.* 36, 105–110, 1998.
9. For introduction to this area see Gould, E., McEwen, B., Tanapat, P., Galea, L.A.M., and Fuchs, E., *J. Neurosci.* 17, 2492–2498, 1997; Cameron, H.A., Woolley, C.S., McEwen, B.S., and Gould, E., *Neurosci.* 56, 337–344, 1993; Gould, E., Reeves, A.J., Fallah, M., Tanapat, P., Gross, C.G., and Fuchs, E., *PNAS USA* 96, 5263–5267, 1999.
10. For an overview integrating basic laboratory experiments and clinical implications, see Goldman, S.A., *J. Neurobiol.* 36, 276–286, 1998. For studies with human brain neurons, see Kirschenbaum, B., Nedergaard, M., Preuss, M., Barami, K., Fraser, R.A.R., and Goldman, S.A., *Cerebral Cortex* 4, 576–589, 1994.
11. Gage, F.H., Bjorklund, A., Stenevi, U., Dunnet, S.B., and Kelly, P.A.T., *Science* 225, 533–536, 1984.
12. Freed, W., Morihisa, J., Spoor, E., Hoffer, B., Olson, L., Seiger, A., and Wyatt, R., *Nature* 292, 351- 352, 1981; Olson, L. in *Synaptic Plasticity and Remodelling*, pp. 485–505, Guilford Press, New York, 1985.

Notes

Chapter 7

1. Virchow, R., *Allg. Z. Psychiatrie* 3, 242–250, 1846.
2. Somjen, G.G., *GLIA* 1, 2–9, 1988.
3. Deiters, O., *Untersuchungen über Gehirn und Ruckenmark des Menschen und der Saugethiere*, M. Schultze, ed, F. Vieweg u. Sohn, Braunschweig, 318 pp., 1865.
4. Ramón y Cajal, S., *Histologie du Systeme Nerveux de l'Homme et des Vertebres*, Paris, vol. 1, 986 pp., reprinted 1952, Cons. Sup. Invest. Cientificas, Inst. Ramón y Cajal, Madrid. Partial translation by J. de la Torre and W.C. Gibson, 1984, *The Neuron and the Glial Cell*, C.C. Thomas, Springfield, 355 pp, 1909, 1911; Contribución al conocimento de la neuroglia del cerebro humano, *Trab. Lab. Invest. Biol. Univ. Madrid* 11, 255–315, 1913.
5. Del Rio Hortega, P., *Trab. Lab. Invest. Biol. Madrid* 18, 37–82, 1920.
6. Penfield, W., *Brain* 47, 430–452, 1924.
7. Lugaro, E., Riv. *Pat. Nerv. Ment.* 12, 225–233, 1907. Source material and translation generously provided by Professor Piergiorgio Strata, University of Turin.
8. Pfrieger, F.W. and Barres, B.A., *Science* 277, 1684–1687, 1997.
9. For examples, see Wu, H., Qu, P., Black, I.B., and Dreyfus, C., *Soc. Neurosci. Abstr.* 20, 1994; Wu, H., Qu, P., Friedman, W., Black, I.B., and Dreyfus, C., *Soc. Neurosci Abstr.* 21, 1995; Dai, Z., Lercher, L.D., Qu, P., Wu, H., Black, I.B., and Dreyfus, C., *Soc. Neurosci.* 22, 1996.

Chapter 8

1. For a classic work on the history of the art of memory, see Yates, F.A., *The Art of Memory*, The University of Chicago Press, Chicago, and Routledge and Kegan Paul Ltd., London, 1966. The story of Simonides is described on pp. 1–2.
2. Cicero, *De Oratore*, 11, 86, 351–354, quoted in Yates, F.A., *The Art of Memory*.
3. Cicero, *De Oratore*, 37, 357, quoted in Yates, F.A., *The Art of Memory*.
4. *Ad Herenium*, 111, 22, quoted in Yates, F.A., *The Art of Memory*.
5. *Marcus Annaeus Seneca Controversiarum Libri*, Lib. 1 Praef. 2, in Yates, F.A. *The Art of Memory*.
6. Augustine, *De anima*, lib. IV. cap. vii, in Yates, F.A., *The Art of Memory*.
7. Ibid., 432, 17, in *The Art of Memory*.
8. Ibid., 432, 9 in *The Art of Memory*.
9. Yates, *The Art of Memory*, p. 34.
10. W.D. Ross, *Aristotle*, London, 1949, p. 144; Ross, *Purva Naturalia*, Oxford, p. 245, 1955.
11. Aristotle, *De memoria et reminiscentia*, translated by W.S. Hett, in Loeb Classical Library, 1935, 452, 8–16.
12. Ibid., 449, 31.
13. Cicero, *Tusculan Disputations* I, 34, 59, quoted from W.C. Wright, Loeb Classical Library in *The Art of Memory*, p. 44.

14. Ibid., I, 25, pp. 62–64.
15. Ibid., I, 25, p. 65.
16. Meyer, V. and Yates, A.J., *J. Neurol., Neurosurg. And Psych.* 18, 44–52, 1955; Milner, B., *Res. Pub. Assoc. Res. Nerv. Ment. Disorders* 36, 244–257, 1958.
17. Kimura, D., *Arch. Neurol.* 8, 264–271, 1963; Prisko, L., Short-term memory in focal cerebral damage. Ph.D. Thesis, McGill University, 1963; Milner, B., *Neuropsych.* 6, 191–210, 1968; Shankweiler, D., Defects in recognition and reproduction of familiar tunes after unilateral temporal lobectomy, Presented at 37th Annual Meeting of the Eastern Psychological Association, New York, April, 1966; Warrington, E. and James, M., *Cortex* 3, 317–326, 1967.
18. Corkin, S. *Neuropsychol.* 3, 339–351, 1965; Milner, B. *Neuropsych.* 3, 317–338, 1965.
19. Penfield, W. and Milner, B., *AMA Arch. Neurol. Psych.* 79, 475–497, 1958.
20. Milner, B., "Memory and the Medial Temporal Regions of the Brain," in *Biological Bases of Memory*, pp. 29–50, K.H. Pribram and D.E. Broadbent, eds., Academic Press, New York,1970.
21. Ibid., p. 37.
22. Ibid., p. 37.
23. Prisko, L., Short-term memory in focal cerebral damage. Unpublished Ph.D. thesis, McGill University, Montreal, Canada, 1963; described in Milner, B. (reference 20).
24. Sidman, M., Stoddard, L.T., and Mohr, J.P., *Neuropsych.* 6, 245–254, 1968.
25. Milner, D., Corkin, S. and Teuber, H.-L., *Neuropsych.* 6, 267–282, 1968.
26. See reference 23, p. 47.
27. Hebb, D.O., *The Organization of Behavior*, Wiley, New York, 1949.
28. Lomo, T., *Acta Physiol. Scand.* 68, suppl. 277, 128, 1966.
29. Bliss, T.V.P. and Lomo, T., *J. Physiol.* 232, 331–356, 1973.
30. Dichter, M.A., Tischler, A.S., and Greene, L.A., *Nature* 268, 501–504, 1977.

Chapter 9

1. The Bhagavad Gita, Juan Mascaro, translator, Penguin Books Ltd., Harmondsworth, Middlesex, U.K., 1962.
2. Kierkegaard, S. *Sickness Unto Death.*
3. See reference 1.
4. Ibid.
5. For review, see Olney, J.W., Excitotoxicity and Neuropsychiatric Disorders, in *Glutamate, Cell Death and Memory*, pp. 77–101, Ascher, P., Choi, D.W., and Christen, Y., eds., Springer-Verlag, 1991.
6. Olney, J.W., *Neurobehav. Toxicol. Teratol.* 6, 455–462, 1984; Olney, J.W., Food Additives, Excitotoxic, in *Encyclopedia of Neuroscience*, Adelman, G., ed., pp. 436–438, Boston, Birkhauser, 1987.
7. Teitelbaum, J.S., Zatorre, R.J., Carpenter, S., Gendron, D., Evans, A.C., Gjedde, A., and Cashman, N.R., *NEJM* 322, 1781–1787, 1990; Perl, T.M.,

Bedard, L., Kosatsky, T., Hockin, J.C., Todd, E.C.D., and Remis, R.S., *NEJM* 322, 1775–1780, 1990.

8. Spencer, P.S., Nunn, P.B., Hugon, J., Ludolph, A.C., Ross, S.M., Roy, D.N., and Robertson, R.C., *Science* 237, 517–522, 1987.

9. Rothman, S., *J. Neurosci.* 4, 1884–1891, 1984.

10. For overview, see Choi, D.W., *J. Neurosci.* 10, 2493–2501, 1990.

11. Mudd, S.H., Irreverre, F., and Laster, L., *Science* 156, 1599–1602, 1967.

12. Plaitakis, A., Beil, S., and Yahr, M., *Ann. Neurol.* 15, 144–153, 1984.

13. Olney, J.W., Misra, C.H., and Rhee, V., *Nature* 264, 659–661, 1976; Spencer, P.S., Schaumburg, H.H., Cohn, D.F., and Seth, P.K., Lathyrism: a useful model of primary lateral sclerosis, in *Research Progress in Motor Neurone Disease*, Rose, F.C., ed., pp. 312–327, Pitman, London, 1984.

Chapter 10

1. Hughes, A., *Int. Rev. Cytol.* 1, 1–7, 1952.

2. Edds, M.V., *Quart. Rev. Biol.* 28, 260–276, 1953.

3. Liu, C.N. and Chambers, W.W., *Arch. Neurol. Psychiat.* 79, 46–61, 1958.

4. Goodman, D.C. and Horel, J. A., *J. Comp. Neurol.* 127, 71–88, 1966.

5. Raisman, G., *Br. Res.* 14, 25–48, 1969.

6. Guth, L. and Windle, W.F., *Exp. Neurol. Suppl.* 5, 1–43, 1970.

7. Bjorklund, A., Stenevi, U., and Svendgaard, N-A., *Nature* 262, 787–790, 1976.

8. Olson, L. and Seiger, A., *Z. Zellforsch.* 135, 175–194, 1972.

9. For an introduction to early work see Bjorklund, A., Stenevi, U., Schmidt, R.H., Dunnett, S.B., and Gage, F.H., *Acta Phys. Scand. Supp.* 522, 11–22, 1983; Bjorklund, A., *Cell Tiss. Res.* 185, 289–302, 1977; Stenevi, U., Bjorklund, A., and Svengaard, N-A., *Br. Res.* 114, 1–20, 1976; Nygren, L.-G., Olson, L., and Seiger, A., *Br. Res.* 129, 227–235, 1977; Stromberg, I., Herren-Marschitz, M., Ungerstedt, U., Ebendahl, T., and Olson L., *Exp. Br. Res.* 60, 335–349, 1985.

10. Gage, F.H., Bjorklund, A., Stenevi, U., Dunnet, S.B., and Kelly, P.A.T., *Science* 225, 533–536, 1984.

11. Benfey, M. and Aguayo, A.J., *Nature* 296, 150–152, 1982.

Afterword

1. Carlsson, A., Lindquist, M., Magnusson, T., and Waldeck, B., *Science* 127, 471, 1958; Cotzias, G.C., Papavasiliou, P.S., and Gelene, R., *NEJM* 280, 337–345, 1969; Birkmayer, W. and Hornykiewicz, O., *Wien. Klin. Wschr.* 73, 787–788, 1961; Ehringer, H. and Hornykiewicz, O., *Wien. Klin. Wschr.* 38, 1236–1239, 1960.

2. Bjorklund, A. and Stenevi, U., *Br. Res.* 177, 555–560, 1979; Bjorklund A., Stenevi, U., Dunnett, S.B., and Iversen, S.D., *Nature* 289, 497–499, 1981; Freed, W.J., Perlow, M.J., Karoum, F., Seiger, A., Olson, L., Hoffer, B.J.,

and Wyatt, R.J., *Ann. Neurol.* 8, 510–518. For review see Olson, L., On the use of transplants to counteract the symptoms of Parkinson's disease: background, experimental models and possible clinical applications, in *Synaptic Plasticity and Remodelling*, Cotman, C., ed., pp. 485–505, Guilford Press, New York, 1985.

3. Lin, L.F., Dohery, D.H., Lile, J.D., Bektesh, S., and Collins, F., *Science* 260, 1130–1132, 1993; Hoffer, B.J., Hoffman, A., Bowenkamp, K., Huettl, P., Hudson, D., Lin, L.F., and Gerhardt, G.A., *Neurosci. Lett.* 182, 107–111, 1994; Hudson, J., Granholm, A.C., Gerhardt, G.A., Henry, M.A., Hoffman, A., Biddle, P., Leela, N.S., Mackerlova, L., Lile, J.D., Collins, F., Hoffer, B.J., *Br. Res. Bull.* 36, 425–432, 1995.

4. For overview see Martinez-Serrano, A. and Bjorklund, A., *TINS* 20, 530–538 (in particular, see pp. 536–537), 1997.

5. Polymeropoulos, M.H., Lavedan, C., Leroy, E., Ide, S.E., Dehejia, A., Dutra, A., Pike, B., Root, H., Rubenstein, J., Boyer, R., Stenroos, E.S., Chandrasekharappa, S.W., Athanassiadou, A., Papatropoulos, T., Johnson, W.G., Lazzarini, A.M., Duvoisin, R., Di Iorio, G., Golbe, L.I., and Nussbaum, R.L., *Science* 276, 2045–2047,1997.

6. See chapters 6 and 10 and references therein.

7. Kempermann, G., Kuhn, H.G., and Gage, F.H., *Nature* 386, 493–495, 1997; Kempermann, G., Kuhn, H., Georg, H., and Gage, F.H., *J. Neurosci.* 18, 3206–3212, 1998; van Praag, H., Kempermann, G., and Gage, F.H., *Nature Neurosci.* 2, 266–270, 1999.

8. For example, see Caterina, M.J., Schumacher, M.A., Tominaga, M.A., Rosen, T.A., Levine, J.D., and Julius, D., *Nature* 389, 816- 824, 1997; Caterina, M.J., Rosen, T.A., Tominaga, M., Brake, A.J., and Julius, D., *Nature* 398, 436- 441, 1999.

9. For introduction and overview, see Kelsey, J.E. and Nemeroff, C.B., Fluoxetine, in *Kaplan & Sadock's Comprehensive Textbook of Psychiatry*, Sadock, B.J. and Sadock, V.A., eds., pp. 2438–2444, Lippincott Williams & Wilkins, Philadelphia, 1999.

10. Schnell, L., Schneider, R., Kolbeck, R., Barde, Y-A., Schwab, M., *Nature* 367, 170–173, 1994; Bregman, B., Kunkel-Bagden, E, Schnell, L., Ning Dal, H., Gao, D., and Schwab, M.E., *Nature* 378, 498–501, 1995.

11. For introduction and review of this exciting, emerging area, see Wickelgren, I., Research news: Teaching the spinal cord to walk, *Science* 279, 319–321, 1998.

12. For examples of PET and functional MRI brain imaging studies and analysis, see Fiez, J.A., Raife, E.A., Balota, D.A., Schwarz, J.P., Raichle, M.E., and Petersen, S.E., *J. Neurosci.* 16, 808–822, 1996; Shulman, G.L., Corbetta, M., Fiez, J.A., Buckner, R.L., Miezen, F.M., Raichle, M.E., and Petersen, S.E., *Human Brain Mapp.* 5, 317–322, John Wiley & Sons, New York, 1997.

INDEX

Index

ABOUT THE AUTHOR

Ira Black has had a passion for the natural world since childhood. He is now a clinical neurologist and neuroscientist who is studying brain plasticity and the regulation of brain and spinal cord genes encoding growth factors, survival factors, and neurotransmitters. He is presently Professor and Chairman, Department of Neuroscience and Cell Biology, Robert Wood Johnson Medical School-UMDNJ, Director of the Joint Graduate Program in Physiology and Neurobiology of Rutgers University and Robert Wood Johnson Medical School, and past president of the Society for Neuroscience of North America. He graduated from the Bronx High School of Science, Columbia College, and Harvard Medical School. He has served on numerous international advisory committees, and is presently on the boards of the Center for Advanced Biotechnology and Medicine of New Jersey, the Christopher Reeve Paralysis Foundation, and the Cure Autism Now Foundation. He has received a number of awards, including the Jacob Javits Award of NIH, the Viktor Hamburger Prize, and the Levi-Montalcini Award. He is the author of approximately 200 scientific articles and of two books, *Cellular and Molecular Biology of Neuronal Development* and *Information in the Brain: A Molecular Perspective*.

When not in the Garden State, he often can be found in the northern tier of the Catskill Mountains, the deserts and canyons of the southwestern United States, or the rainforests of Central America.